CONSTRUCTION CRAFTS CORE UNITS

LEVEL 2 DIPLOMA

Leeds
College of
Building

Nelson Thornes

Text © Nelson Thornes 2013

Original illustrations © Nelson Thornes Ltd 2013

The rights of Jon Sutherland, Diane Canwell, Althea Brooks and Leeds College of Building to be identified as authors of this work has been asserted by them in accordance with the Copyright, Designs and Patents Act 1988.

Published in 2013 by:
Nelson Thornes Ltd
Delta Place
27 Bath Road
CHELTENHAM
GL53 7TH
United Kingdom

13 14 15 16 17 / 10 9 8 7 6 5 4 3 2 1

A catalogue record for this book is available from the British Library

ISBN 978 1 4085 2312 4

Cover photograph: Fotolia

Page make-up by GreenGate Publishing Services, Tonbridge, Kent

Printed in China by 1010 Printing International Ltd

Note to learners and tutors

This book clearly states that a risk assessment should be undertaken and the correct PPE worn for the particular activities before any practical activity is carried out. Risk assessments were carried out before photographs for this book were taken and the models are wearing the PPE deemed appropriate for the activity and situation. This was correct at the time of going to print. Colleges may prefer that their learners wear additional items of PPE not featured in the photographs in this book and should instruct learners to do so in the standard risk assessments they hold for activities undertaken by their learners. Learners should follow the standard risk assessments provided by their college for each activity they undertake which will determine the PPE they wear.

CONTENTS

INTRODUCTION

About this book

This book has been written to support the core units of the Cskills Awards Level 2 construction Diplomas. It covers all three mandatory units of the diploma, so you can use this book for any of the new construction crafts diplomas.

This book contains a number of features to help you acquire the knowledge you need. We've included additional features to show how the skills and knowledge can be applied to the workplace, as well as tips and advice on how you can improve your chances of gaining employment.

The features include:

 KEY TERMS

DID YOU KNOW?

PRACTICAL TIP

 REED TIP ...

CASE STUDY

TEST YOURSELF

* chapter openers which list the learning outcomes you must achieve in each unit

* key terms that provide explanations of important terminology that you will need to know and understand

* Did you know? margin notes to provide key facts that are helpful to your learning

* practical tips to explain facts or skills to remember when undertaking practical tasks

* Reed tips to offer advice about work, building your CV and how to apply the skills and knowledge you have learnt in the workplace

* case studies that are based on real tradespeople who have undertaken apprenticeships and explain why the skills and knowledge you learn with your training provider are useful in the workplace

* Test yourself multiple choice questions that appear at the end of each unit to give you the chance to revise what you have learnt and to practise your assessment (your tutor will give you the answers to these questions).

Further support for this book can be found at our website,

www.planetvocational.com/subjects/build

Planet Vocational

CONTRIBUTORS TO THIS BOOK

Reed Property & Construction

Reed Property & Construction specialises in placing staff at all levels, in both temporary and permanent positions, across the complete lifecycle of the construction process. Our consultants work with most major construction companies in the UK and our clients are involved with the design, build and maintenance of infrastructure projects throughout the UK.

Expert help

As a leading recruitment consultancy for mid–senior level construction staff in the UK, Reed Property & Construction is ideally placed to advise new workers entering the sector, from building a CV to providing expertise and sharing our extensive sector knowledge with you. That's why, throughout this book, you will find helpful hints from our highly experienced consultants, all designed to help you find that first step on the construction career ladder. These tips range from advice on CV writing to interview tips and techniques, and are all linked in with the learning material in this book.

Work-related advice

Reed Property & Construction has gained insights from some of our biggest clients – leading recruiters within the industry – to help you understand the mind-set of potential employers. This includes the traits and skills that they would like to see in their new employees, why you need the skills taught in this book and how they are used on a day to day basis within their organisations.

Getting your first job

This invaluable information is not available anywhere else and is all geared towards helping you gain a position with an employer once you've completed your studies. Entry level positions are not usually offered by recruitment companies, but the advice we've provided will help you to apply for jobs in construction and hopefully gain your first position as a skilled worker.

CONTRIBUTORS TO THIS BOOK

The case studies in this book feature staff from Laing O'Rourke and South Tyneside Homes.

Laing O'Rourke is an international engineering company that constructs large-scale building projects all over the world. Originally formed from two companies, John Laing (founded in 1848) and R O'Rourke and Son (founded in 1978) joined forces in 2001.

At Laing O'Rourke, there is a strong and unique apprenticeship programme. It runs a four-year 'Apprenticeship Plus' scheme in the UK, combining formal college education with on-the-job training. Apprentices receive support and advice from mentors and experienced tradespeople, and are given the option of three different career pathways upon completion: remaining on site, continuing into a further education programme, or progressing into supervision and management.

The company prides itself on its people development, supporting educational initiatives and investing in its employees. Laing O'Rourke believes in collaboration and teamwork as a path to achieving greater success, and strives to maintain exceptionally high standards in workplace health and safety.

South Tyneside Homes

South Tyneside Council's Housing Company

South Tyneside Homes was launched in 2006, and was previously part of South Tyneside Council. It now works in partnership with the council to repair and maintain 18,000 properties within the borough, including delivering parts of the Decent Homes Programme.

South Tyneside Homes believes in putting back into the community, with 90 per cent of its employees living in the borough itself. Equality and diversity, as well as health and wellbeing of staff, is a top priority, and it has achieved the Gold Status Investors in People Award.

South Tyneside Homes is committed to the development of its employees, providing opportunities for further education and training and great career paths within the company – 80 per cent of its management team started as apprentices with the company. As well as looking after its staff and their community, the company looks after the environment too, running a renewable energy scheme for council tenants in order to reduce carbon emissions and save tenants money.

The apprenticeship programme at South Tyneside Homes has been recognised nationally, having trained over 80 young people in five main trade areas over the past six years. One of the UK's Top 100 Apprenticeship Employers, it is an Ambassador on the panel of the National Apprentice Service. It has won the Large Employer of the Year Award at the National Apprenticeship Awards and several of its apprentices have been nominated for awards, including winning the Female Apprentice of the Year for the local authority.

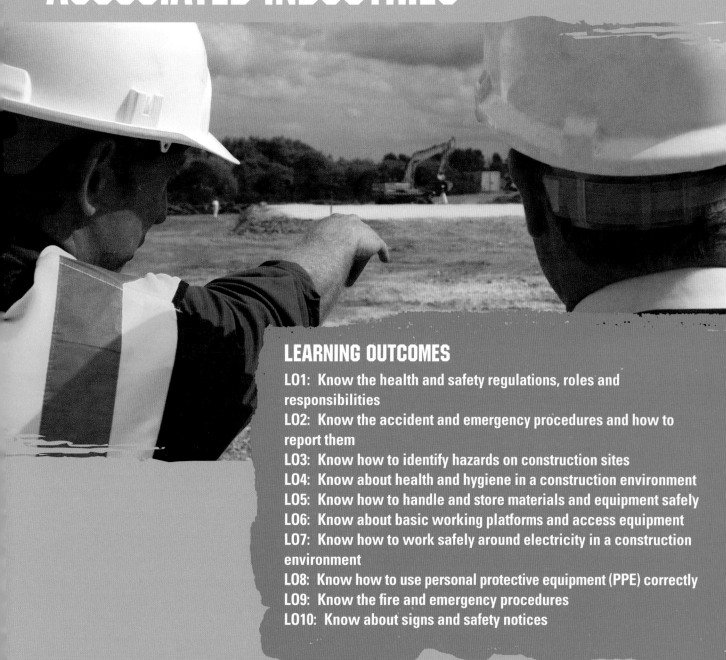

Chapter 1

Unit CSA–L1Core01

HEALTH, SAFETY AND WELFARE IN CONSTRUCTION AND ASSOCIATED INDUSTRIES

LEARNING OUTCOMES

LO1: Know the health and safety regulations, roles and responsibilities

LO2: Know the accident and emergency procedures and how to report them

LO3: Know how to identify hazards on construction sites

LO4: Know about health and hygiene in a construction environment

LO5: Know how to handle and store materials and equipment safely

LO6: Know about basic working platforms and access equipment

LO7: Know how to work safely around electricity in a construction environment

LO8: Know how to use personal protective equipment (PPE) correctly

LO9: Know the fire and emergency procedures

LO10: Know about signs and safety notices

INTRODUCTION

The aim of this chapter is to:

* help you to source relevant safety information
* help you to use the relevant safety procedures at work.

KEY TERMS

HASAWA

– the Health and Safety at Work etc. Act outlines your and your employer's health and safety responsibilities.

COSHH

– the Control of Substances Hazardous to Health Regulations are concerned with controlling exposure to hazardous materials.

DID YOU KNOW?

In 2011 to 2012, there were 49 fatal accidents in the construction industry in the UK. (*Source* HSE, www.hse.gov.uk)

KEY TERMS

HSE

– the Health and Safety Executive, which ensures that health and safety laws are followed.

Accident book

– this is required by law under the Social Security (Claims and Payments) Regulations 1979. Even minor accidents need to be recorded by the employer. For the purposes of RIDDOR, hard copy accident books or online records of incidents are equally acceptable.

HEALTH AND SAFETY REGULATIONS, ROLES AND RESPONSIBILITIES

The construction industry can be dangerous, so keeping safe and healthy at work is very important. If you are not careful, you could injure yourself in an accident or perhaps use equipment or materials that could damage your health. Keeping safe and healthy will help ensure that you have a long and injury-free career.

Although the construction industry is much safer today than in the past, more than 2,000 people are injured and around 50 are killed on site every year. Many others suffer from long-term ill-health such as deafness, spinal damage, skin conditions or breathing problems.

Key health and safety legislation

Laws have been created in the UK to try to ensure safety at work. Ignoring the rules can mean injury or damage to health. It can also mean losing your job or being taken to court.

The two main laws are the Health and Safety at Work etc. Act **(HASAWA)** and the Control of Substances Hazardous to Health Regulations **(COSHH)**.

The Health and Safety at Work etc. Act (HASAWA) (1974)

This law applies to all working environments and to all types of worker, sub-contractor, employer and all visitors to the workplace. It places a duty on everyone to follow rules in order to ensure health, safety and welfare. Businesses must manage health and safety risks, for example by providing appropriate training and facilities. The Act also covers first aid, accidents and ill health.

Reporting of Injuries, Diseases and Dangerous Occurrences Regulations (RIDDOR) (1995)

Under RIDDOR, employers are required to report any injuries, diseases or dangerous occurrences to the **Health and Safety Executive (HSE)**. The regulations also state the need to maintain an **accident book**.

Control of Substances Hazardous to Health (COSHH) (2002)

In construction, it is common to be exposed to substances that could cause ill health. For example, you may use oil-based paints or preservatives, or work in conditions where there is dust or bacteria.

Employers need to protect their employees from the risks associated with using hazardous substances. This means assessing the risks and deciding on the necessary precautions to take.

Any control measures (things that are being done to reduce the risk of people being hurt or becoming ill) have to be introduced into the workplace and maintained; this includes monitoring an employee's exposure to harmful substances. The employer will need to carry out health checks and ensure that employees are made aware of the dangers and are supervised.

Control of Asbestos at Work Regulations (2012)

Asbestos was a popular building material in the past because it was a good insulator, had good fire protection properties and also protected metals against corrosion. Any building that was constructed before 2000 is likely to have some asbestos. It can be found in pipe insulation, boilers and ceiling tiles. There is also asbestos cement roof sheeting and there is a small amount of asbestos in decorative coatings such as Artex.

Asbestos has been linked with lung cancer, other damage to the lungs and breathing problems. The regulations require you and your employer to take care when dealing with asbestos:

* You should always assume that materials contain asbestos unless it is obvious that they do not.

* A record of the location and condition of asbestos should be kept.

* A risk assessment should be carried out if there is a chance that anyone will be exposed to asbestos.

The general advice is as follows:

* Do not remove the asbestos. It is not a hazard unless it is removed or damaged.

* Remember that not all asbestos presents the same risk. Asbestos cement is less dangerous than pipe insulation.

* Call in a specialist if you are uncertain.

Provision and Use of Work Equipment Regulations (PUWER) (1998)

PUWER concerns health and safety risks related to equipment used at work. It states that any risks arising from the use of equipment must either be prevented or controlled, and all suitable safety measures must have been taken. In addition, tools need to be:

* suitable for their intended use

* safe

> **REED TIP**
>
> Employers will want to know that you understand the importance of health and safety. Make sure you know the reasons for each safe working practice.

* well maintained

* used only by those who have been trained to do so.

Manual Handling Operations Regulations (1992)

These regulations try to control the risk of injury when lifting or handling bulky or heavy equipment and materials. The regulations state as follows:

* Hazardous manual handling should be avoided if possible.

* An assessment of hazardous manual handling should be made to try to find alternatives.

* You should use mechanical assistance where possible.

* The main idea is to look at how manual handling is carried out and finding safer ways of doing it.

Personal Protection at Work Regulations (PPE) (1992)

This law states that employers must provide employees with personal protective equipment **(PPE)** at work whenever there is a risk to health and safety. PPE needs to be:

* suitable for the work being done

* well maintained and replaced if damaged

* properly stored

* correctly used (which means employees need to be trained in how to use the PPE properly).

Work at Height Regulations (2005)

Whenever a person works at any height there is a risk that they could fall and injure themselves. The regulations place a duty on employers or anyone who controls the work of others. This means that they need to:

* plan and organise the work

* make sure those working at height are **competent**

* assess the risks and provide appropriate equipment

* manage work near or on fragile surfaces

* ensure equipment is inspected and maintained.

In all cases the regulations suggest that, if it is possible, work at height should be avoided. Perhaps the job could be done from ground level? If it is not possible, then equipment and other measures are needed to prevent the risk of falling. When working at height measures also need to be put in place to minimise the distance someone might fall.

KEY TERMS

PPE

– personal protective equipment can include gloves, goggles and hard hats.

Competent

– to be competent an organisation or individual must have:

* sufficient knowledge of the tasks to be undertaken and the risks involved

* the experience and ability to carry out their duties in relation to the project, to recognise their limitations and take appropriate action to prevent harm to those carrying out construction work, or those affected by the work.

(*Source* HSE)

Figure 1.1 Examples of personal protective equipment

Employer responsibilities under HASAWA

HASAWA states that employers with five or more staff need their own health and safety policy. Employers must assess any risks that may be involved in their workplace and then introduce controls to reduce these risks. These risk assessments need to be reviewed regularly.

Employers also need to supply personal protective equipment (PPE) to all employees when it is needed and to ensure that it is worn when required.

Specific employer responsibilities are outlined in Table 1.1.

Employee responsibilities under HASAWA

HASAWA states that all those operating in the workplace must aim to work in a safe way. For example, they must wear any PPE provided and look after their equipment. Employees should not be charged for PPE or any actions that the employer needs to take to ensure safety.

Specific employer responsibilities are outlined in Table 1.1. Table 1.2 identifies the key employee responsibilities.

KEY TERMS

Risk

– the likelihood that a person may be harmed if they are exposed to a hazard.

Hazard

– a potential source of harm, injury or ill-health.

Near miss

– any incident, accident or emergency that did not result in an injury but could have done so.

Employer responsibility	Explanation
Safe working environment	Where possible all potential risks and hazards should be eliminated.
Adequate staff training	When new employees begin a job their induction should cover health and safety. There should be ongoing training for existing employees on risks and control measures.
Health and safety information	Relevant information related to health and safety should be available for employees to read and have their own copies.
Risk assessment	Each task or job should be investigated and potential risks identified so that measures can be put in place. A risk assessment and method statement should be produced. The method statement will tell you how to carry out the task, what PPE to wear, equipment to use and the sequence of its use.
Supervision	A competent and experienced individual should always be available to help ensure that health and safety problems are avoided.

Table 1.1 Employer responsibilities under HASAWA

Employee responsibility	Explanation
Working safely	Employees should take care of themselves, only do work that they are competent to carry out and remove obvious hazards if they are seen.
Working in partnership with the employer	Co-operation is important and you should never interfere with or misuse any health and safety signs or equipment. You should always follow the site rules.
Reporting hazards, near misses and accidents correctly	Any health and safety problems should be reported and discussed, particularly a near miss or an actual accident.

Table 1.2 Employee responsibilities under HASAWA

Health and Safety Executive

The Health and Safety Executive (HSE) is responsible for health, safety and welfare. It carries out spot checks on different workplaces to make sure that the law is being followed.

HSE inspectors have access to all areas of a construction site and can also bring in the police. If they find a problem then they can issue an **improvement notice**. This gives the employer a limited amount of time to put things right.

In serious cases, the HSE can issue a **prohibition notice**. This means all work has to stop until the problem is dealt with. An employer, the employees or **sub-contractors** could be taken to court.

The roles and responsibilities of the HSE are outlined in Table 1.3.

Responsibility	Explanation
Enforcement	It is the HSE's responsibility to reduce work-related death, injury and ill health. It will use the law against those who put others at risk.
Legislation and advice	The HSE will use health and safety legislation to serve improvement or prohibition notices or even to prosecute those who break health and safety rules. Inspectors will provide advice either face-to-face or in writing on health and safety matters.
Inspection	The HSE will look at site conditions, standards and practices and inspect documents to make sure that businesses and individuals are complying with health and safety law.

Table 1.3 HSE roles and responsibilities

Sources of health and safety information

There is a wide variety of health and safety information. Most of it is available free of charge, while other organisations may make a charge to provide information and advice. Table 1.4 outlines the key sources of health and safety information.

Source	Types of information	Website
Health and Safety Executive (HSE)	The HSE is the primary source of work-related health and safety information. It covers all possible topics and industries.	www.hse.gov.uk
Construction Industry Training Board (CITB)	The national training organisation provides key information on legislation and site safety.	www.citb.co.uk
British Standards Institute (BSI)	Provides guidelines for risk management, PPE, fire hazards and many other health and safety-related areas.	www.bsigroup.com
Royal Society for the Prevention of Accidents (RoSPA)	Provides training, consultancy and advice on a wide range of health and safety issues that are aimed to reduce work related accidents and ill health.	www.rospa.com
Royal Society for Public Health (RSPH)	Has a range of qualifications and training programmes focusing on health and safety.	www.rsph.org.uk

Table 1.4 Health and safety information

Informing the HSE

The HSE requires the reporting of:

* deaths and injuries – any **major injury**, **over 7-day injury** or death

* occupational disease

* dangerous occurrence – a collapse, explosion, fire or collision

* gas accidents – any accidental leaks or other incident related to gas.

KEY TERMS

Major injury

– any fractures, amputations, dislocations, loss of sight or other severe injury.

Over 7-day injury

– an injury that has kept someone off work for more than seven days.

Enforcing guidance

Work-related injuries and illnesses affect huge numbers of people. According to the HSE, 1.1 million working people in the UK suffered from a work-related illness in 2011 to 2012. Across all industries, 173 workers were killed, 111,000 other injuries were reported and 27 million working days were lost.

The construction industry is a high risk one and, although only around 5 per cent of the working population is in construction, it accounts for 10 per cent of all major injuries and 22 per cent of fatal injuries.

The good news is that enforcing guidance on health and safety has driven down the numbers of injuries and deaths in the industry. Only 20 years ago over 120 construction workers died in workplace accidents each year. This is now reduced to fewer than 60 a year.

However, there is still more work to be done and it is vital that organisations such as the HSE continue to enforce health and safety and continue to reduce risks in the industry.

DID YOU KNOW?

Workplace injuries cost the UK £13.4bn in 2010 to 2011.

On-site safety inductions and toolbox talks

The HSE suggests that all new workers arriving on site should attend a short induction session on health and safety. It should:

* show the commitment of the company to health and safety

* explain the health and safety policy

* explain the roles individuals play in the policy

* state that each individual has a legal duty to contribute to safe working

* cover issues like excavations, work at height, electricity and fire risk

* provide a layout of the site and show evacuation routes

* identify where fire fighting equipment is located

* ensure that all employees have evidence of their skills

* stress the importance of signing in and out of the site.

DID YOU KNOW?

Toolbox talks are normally given by a supervisor and often take place on site, either during the course of a normal working day or when someone has been seen working in an unsafe way. CITB produces a book called *GT700 Toolbox Talks* which covers a range of health and safety topics, from trying a new process and using new equipment to particular hazards or work practices.

Behaviour and actions that could affect others

It is the responsibility of everyone on site not only to look after their own health and safety, but also to ensure that their actions do not put anyone else at risk.

Trying to carry out work that you are not competent to do is not only dangerous to yourself but could compromise the safety of others.

Simple actions, such as ensuring that all of your rubbish and waste is properly disposed of, will go a long way to removing hazards on site that could affect others.

Just as you should not create a hazard, ignoring an obvious one is just as dangerous. You should always obey site rules and particularly the health and safety rules. You should follow any instructions you are given.

PRACTICAL TIP

If you come across any health and safety problems you should report them so that they can be controlled.

ACCIDENT AND EMERGENCY PROCEDURES

All sites will have specific procedures for dealing with accidents and emergencies. An emergency will often mean that the site needs to be evacuated, so you should know in advance where to assemble and who to report to. The site should never be re-entered without authorisation from an individual in charge or the emergency services.

Types of emergencies

Emergencies are incidents that require immediate action. They can include:

* fires
* spillages or leaks of chemicals or other hazardous substances, such as gas
* failure of a scaffold
* collapse of a wall or trench
* a health problem
* an injury
* bombs and security alerts.

Legislation and reporting accidents

RIDDOR (1995) puts a duty on employers, anyone who is self-employed, or an individual in control of the work, to report any serious workplace accidents, occupational diseases or dangerous occurrences (also known as near misses).

The report has to be made by these individuals and, if it is serious enough, the responsible person may have to fill out a RIDDOR report.

Figure 1.2 It's important that you know where your company's fire-fighting equipment is located

Injuries, diseases and dangerous occurrences

Construction sites can be dangerous places, as we have seen. The HSE maintains a list of all possible injuries, diseases and dangerous occurrences, particularly those that need to be reported.

Injuries

There are two main classifications of injuries: minor and major. A minor injury can usually be handled by a competent first aider, although it is often a good idea to refer the individual to their doctor or to the hospital. Typical minor injuries can include:

* minor cuts
* minor burns
* exposure to fumes.

Major injuries are more dangerous and will usually require the presence of an ambulance with paramedics. Major injuries can include:

* bone fracture
* concussion
* unconsciousness
* electric shock.

Diseases

There are several different diseases and health issues that have to be reported, particularly if a doctor notifies that a disease has been diagnosed. These include:

* poisoning
* infections
* skin diseases
* occupational cancer
* lung diseases
* hand/arm vibration syndrome.

Dangerous occurrences

Even if something happens that does not result in an injury, but could easily have done so, it is classed as a dangerous occurrence. It needs to be reported immediately and then followed up by an accident report form. Dangerous occurrences can include:

* accidental release of a substance that could damage health

* anything coming into contact with overhead power lines

* an electrical problem that caused a fire or explosion

* collapse or partial collapse of scaffolding over 5m high.

PRACTICAL TIP

An up-to-date list of dangerous occurrences is maintained by the Health and Safety Executive.

Recording accidents and emergencies

The Reporting of Injuries, Diseases and Dangerous Occurrences Regulations (RIDDOR) (1995) requires employers to:

* report any relevant injuries, diseases or dangerous occurrences to the Health and Safety Executive (HSE)

* keep records of incidents in a formal and organised manner (for example, in an accident book or online database).

After an accident, you may need to complete an accident report form – either in writing or online. This form may be completed by the person who was injured or the first aider.

On the accident report form you need to note down:

* the casualty's personal details, e.g. name, address, occupation
* the name of the person filling in the report form
* the details of the accident.

In addition, the person reporting the accident will need to sign the form.

On site a trained first aider will be the first individual to try and deal with the situation. In addition to trying to save life, stop the condition from getting worse and getting help, they will also record the occurrence.

On larger sites there will be a health and safety officer, who would keep records and documentation detailing any accidents and emergencies that have taken place on site. All companies should keep such records; it may be a legal requirement for them to do so under RIDDOR and it is good practice to do so in case the HSE asks to see it.

Importance of reporting accidents and near misses

Reporting incidents is not just about complying with the law or providing information for statistics. Each time an accident or near miss takes place it means lessons can be learned and future problems avoided.

The accident or near miss can alert the business or organisation to a potential problem. They can then take steps to ensure that it does not occur in the future.

Major and minor injuries and near misses

RIDDOR defines a major injury as:

* a fracture (but not to a finger, thumb or toes)
* a dislocation
* an amputation
* a loss of sight in an eye
* a chemical or hot metal burn to the eye
* a penetrating injury to the eye
* an electric shock or electric burn leading to unconsciousness and/or requiring resuscitation
* hyperthermia, heat-induced illness or unconsciousness
* asphyxia
* exposure to a harmful substance
* inhalation of a substance
* acute illness after exposure to toxins or infected materials.

A minor injury could be considered as any occurrence that does not fall into any of the above categories.

A near miss is any incident that did not actually result in an injury but which could have caused a major injury if it had done so. Non-reportable near misses are useful to record as they can help to identify potential problems. Looking at a list of near misses might show patterns for potential risk.

Accident trends

We have already seen that the HSE maintains statistics on the number and types of construction accidents. The following are among the 2011/2012 construction statistics:

* There were 49 fatalities.

* There were 5,000 occupational cancer patients.

* There were 74,000 cases of work-related ill health.

* The most common types of injury were caused by falls, although many injuries were caused by falling objects, collapses and electricity. A number of construction workers were also hurt when they slipped or tripped, or were injured while lifting heavy objects.

Accidents, emergencies and the employer

Even less serious accidents and injuries can cost a business a great deal of money. But there are other costs too:

* Poor company image – if a business does not have health and safety controls in place then it may get a reputation for not caring about its employees. The number of accidents and injuries may be far higher than average.

* Loss of production – the injured individual might have to be treated and then may need a period of time off work to recover. The loss of production can include those who have to take time out from working to help the injured person and the time of a manager or supervisor who has to deal with all the paperwork and problems.

* Insurance – each time there is an accident or injury claim against the company's insurance the premiums will go up. If there are many accidents and injuries the business may find it impossible to get insurance. It is a legal requirement for a business to have insurance so in the end that company might have to close down.

* Closure of site – if there is a serious accident or injury then the site may have to be closed while investigations take place to discover the reason, or who was responsible. This could cause serious delays and loss of income for workers and the business.

Accident and emergency authorised personnel

Several different groups of people could be involved in dealing with accident and emergency situations. These are listed in Table 1.5.

Authorised personnel	Role
First aiders and emergency responders	These are employees on site and in the workforce who have been trained to be the first to respond to accidents and injuries. The minimum provision of an appointed person would be someone who has had basic first aid training. The appointment of a first aider is someone who has attained a higher or specific level of training. A construction site with fewer than 5 employees needs an appointed first aider. A construction site with up to 50 employees requires a trained first aider, and for bigger sites at least one trained first aider is required for every 50 people.
Supervisors and managers	These have the responsibility of managing the site and would have to organise the response and contact emergency services if necessary. They would also ensure that records of any accidents are completed and up to date and notify the HSE if required.
Health and Safety Executive	The HSE requires businesses to investigate all accidents and emergencies. The HSE may send an inspector, or even a team, to investigate and take action if the law has been broken.
Emergency services	Calling the emergency services depends on the seriousness of the accident. Paramedics will take charge of the situation if there is a serious injury and if they feel it necessary will take the individual to hospital.

Table 1.5 People who deal with accident and emergency situations

DID YOU KNOW?

The three main emergency services in the UK are: the Fire Service (for fire and rescue); the Ambulance Service (for medical emergencies); the Police (for an immediate police response). Call them on 999 only if it is an emergency.

The basic first aid kit

BS 8599 relates to first aid kits, but it is not legally binding. The contents of a first aid box will depend on an employer's assessment of their likely needs. The HSE does not have to approve the contents of a first aid box but it states that where the work involves low level hazards the minimum contents of a first aid box should be:

* a copy of its leaflet on first aid – *HSE Basic advice on first aid at work*
* 20 sterile plasters of assorted size
* 2 sterile eye pads
* 4 sterile triangular bandages
* 6 safety pins
* 2 large sterile, unmedicated wound dressings
* 6 medium-sized sterile unmedicated wound dressings
* 1 pair of disposable gloves.

The HSE also recommends that no tablets or medicines are kept in the first aid box.

Figure 1.3 A typical first aid box

What to do if you discover an accident

When an accident happens it may not only injure the person involved directly, but it may also create a hazard that could then injure others. You need to make sure that the area is safe enough for you or someone else to help the injured person. It may be necessary to turn off the electrical supply or remove obstructions to the site of the accident.

The first thing that needs to be done if there is an accident is to raise the alarm. This could mean:

* calling for the first aider

* phoning for the emergency services

* dealing with the problem yourself.

How you respond will depend on the severity of the injury.

You should follow this procedure if you need to contact the emergency services:

* Find a telephone away from the emergency.

* Dial 999.

* You may have to go through a switchboard. Carefully listen to what the operator is saying to you and try to stay calm.

* When asked, give the operator your name and location, and the name of the emergency service or services you require.

* You will then be transferred to the appropriate emergency service, who will ask you questions about the accident and its location. Answer the questions in a clear and calm way.

* Once the call is over, make sure someone is available to help direct the emergency services to the location of the accident.

IDENTIFYING HAZARDS

As we have already seen, construction sites are potentially dangerous places. The most effective way of handling health and safety on a construction site is to spot the hazards and deal with them before they can cause an accident or an injury. This begins with basic housekeeping and carrying out risk assessments. It also means having a procedure in place to report hazards so that they can be dealt with.

Good housekeeping

Work areas should always be clean and tidy. Sites that are messy, strewn with materials, equipment, wires and other hazards can prove to be very dangerous. You should:

* always work in a tidy way

* never block fire exits or emergency escape routes

* never leave nails and screws scattered around

* ensure you clean and sweep up at the end of each working day

* not block walkways

* never overfill skips or bins

* never leave food waste on site.

Risk assessments and method statements

It is a legal requirement for employers to carry out risk assessments. This covers not only those who are actually working on a particular job, but other workers in the immediate area, and others who might be affected by the work.

It is important to remember that when you are carrying out work your actions may affect the safety of other people. It is important, therefore, to know whether there are any potential hazards. Once you know what these hazards are you can do something to either prevent or reduce them as a risk. Every job has potential hazards.

There are five simple steps to carrying out a risk assessment, which are shown in Table 1.6, using the example of repointing brickwork on the front face of a dwelling.

Step	Action	Example
1	Identify hazards	The property is on a street with a narrow pavement. The damaged brickwork and loose mortar need to be removed and placed in a skip below. Scaffolding has been erected. The road is not closed to traffic.
2	Identify who is at risk	The workers repointing are at risk as they are working at height. Pedestrians and vehicles passing are at risk from the positioning of the skip and the chance that debris could fall from height.
3	What is the risk from the hazard that may cause an accident?	The risk to the workers is relatively low as they have PPE and the scaffolding has been correctly erected. The risk to those passing by is higher, as they are unaware of the work being carried out above them.
4	Measures to be taken to reduce the risk	Station someone near the skip to direct pedestrians and vehicles away from the skip while the work is being carried out. Fix a secure barrier to the edge of the scaffolding to reduce the chance of debris falling down. Lower the bricks and mortar debris using a bucket or bag into the skip and not throwing them from the scaffolding. Consider carrying out the work when there are fewer pedestrians and less traffic on the road.
5	Monitor the risk	If there are problems with the first stages of the job, you need to take steps to solve them. If necessary consider taking the debris by hand through the building after removal.

Table 1.6 A five-step risk assessment for repointing brickwork

Your employer should follow these working practices, which can help to prevent accidents or dangerous situations occurring in the workplace:

* *Risk assessments* look carefully at what could cause an individual harm and how to prevent this. This is to ensure that no one should be injured or become ill as a result of their work. Risk assessments identify how likely it is that an accident might happen and the consequences of it happening. A risk factor is worked out and control measures created to try to offset them.

* *Method statements,* however brief, should be available for every risk assessment. They summarise risk assessments and other findings to provide guidance on how the work should be carried out.

* *Permit to work systems* are used for very high risk or even potentially fatal activities. They are checklists that need to be completed before the work begins. They must be signed by a supervisor.

* *A hazard book* lists standard tasks and identifies common hazards. These are useful tools to help quickly identify hazards related to particular tasks.

Types of hazards

Typical construction accidents can include:

* fires and explosions

* slips, trips and falls

* burns, including those from chemicals

* falls from scaffolding, ladders and roofs

* electrocution

* injury from faulty machinery

* power tool accidents

* being hit by construction debris

* falling through holes in flooring

We will look at some of the more common hazards in a little more detail.

Fires

Fires need oxygen, heat and fuel to burn. Even a spark can provide enough heat needed to start a fire, and anything flammable, such as petrol, paper or wood, provides the fuel. It may help to remember the 'triangle of fire' – heat, oxygen and fuel are all needed to make fire so remove one or more to help prevent or stop the fire.

Tripping

Leaving equipment and materials lying around can cause accidents, as can trailing cables and spilt water or oil. Some of these materials are also potential fire hazards.

Chemical spills

If the chemicals are not hazardous then they just need to be mopped up. But sometimes they do involve hazardous materials and there will be an existing plan on how to deal with them. A risk assessment will have been carried out.

Falls from height

A fall even from a low height can cause serious injuries. Precautions need to be taken when working at height to avoid permanent injury. You should also consider falls into open excavations as falls from height. All the same precautions need to be in place to prevent a fall.

Burns

Burns can be caused not only by fires and heat, but also from chemicals and solvents. Electricity and wet concrete and cement can also burn skin. PPE is often the best way to avoid these dangers. Sunburn is a common and uncomfortable form of burning and sunscreen should be made available. For example, keeping skin covered up will help to prevent sunburn. You might think a tan looks good, but it could lead to skin cancer.

Electrical

Electricity is hazardous and electric shocks can cause burns and muscle damage, and can kill.

Exposure to hazardous substances

We look at hazardous substances in more detail on pages 20–1. COSHH regulations identify hazardous substances and require them to be labelled. You should always follow the instructions when using them.

Plant and vehicles

On busy sites there is always a danger from moving vehicles and heavy plant. Although many are fitted with reversing alarms, it may not be easy to hear them over other machinery and equipment. You should always ensure you are not blocking routes or exits. Designated walkways separate site traffic and pedestrians – this includes workers who are walking around the site. Crossing points should be in place for ease of movement on site.

Reporting hazards

We have already seen that hazards have the potential to cause serious accidents and injuries. It is therefore important to report hazards and there are different methods of doing this.

The first major reason to report hazards is to prevent danger to others, whether they are other employees or visitors to the site. It is vital to prevent accidents from taking place and to quickly correct any dangerous situations.

Injuries, diseases and actual accidents all need to be reported and so do dangerous occurrences. These are incidents that do not result in an actual injury, but could easily have hurt someone.

Accidents need to be recorded in an accident book, computer database or other secure recording system, as do near misses. Again it is a legal requirement to keep appropriate records of accidents and every company will have a procedure for this which they should tell you about. Everyone should know where the book is kept or how the records are made. Anyone that has been hurt or has taken part in dealing with an occurrence should complete the details of what has happened. Typically this will require you to fill in:

* the date, time and place of the incident

* how it happened

* what was the cause

* how it was dealt with

* who was involved

* signature and date.

The details in the book have to be transferred onto an official HSE report form.

As far as is possible, the site, company or workplace will have set procedures in place for reporting hazards and accidents. These procedures will usually be found in the place where the accident book or records are stored. The location tends to be posted on the site notice board.

How hazards are created

Construction sites are busy places. There are constantly new stages in development. As each stage is begun a whole new set of potential hazards need to be considered.

At the same time, new workers will always be joining the site. It is mandatory for them to be given health and safety instruction during induction. But sometimes this is impossible due to pressure of work or availability of trainers.

Construction sites can become even more hazardous in times of extreme weather:

* Flooding – long periods of rain can cause trenches to fill with water, cellars to be flooded and smooth surfaces to become extremely wet and slippery.

* Wind – strong winds may prevent all work at height. Scaffolding may have become unstable, unsecured roofing materials may come loose, dry-stored materials such as sand and cement may have been blown across the site.

- Heat – this can change the behaviour of materials: setting quicker, failing to cure and melting. It can also seriously affect the health of the workforce through dehydration and heat exhaustion.

- Snow – this can add enormous weight to roofs and other structures and could cause collapse. Snow can also prevent access or block exits and can mean that simple and routine work becomes impossible due to frozen conditions.

Storing combustibles and chemicals

A combustible substance can be both flammable and explosive. There are some basic suggestions from the HSE about storing these:

- Ventilation – the area should be well ventilated to disperse any vapours that could trigger off an explosion.

- Ignition – an ignition is any spark or flame that could trigger off the vapours, so materials should be stored away from any area that uses electrical equipment or any tool that heats up.

- Containment – the materials should always be kept in proper containers with lids and there should be spillage trays to prevent any leak seeping into other parts of the site.

- Exchange – in many cases it can be possible to find an alternative material that is less dangerous. This option should be taken if possible.

- Separation – always keep flammable substances away from general work areas. If possible they should be partitioned off.

Combustible materials can include a large number of commonly used substances, such as cleaning agents, paints and adhesives.

HEALTH AND HYGIENE

Just as hazards can be a major problem on site, other less obvious problems relating to health and hygiene can also be an issue. It is both your responsibility and that of your employer to make sure that you stay healthy.

The employer will need to provide basic welfare facilities, no matter where you are working and these must have minimum standards.

Welfare facilities

Welfare facilities can include a wide range of different considerations, as can be seen in Table 1.7.

Facilities	Purpose and minimum standards
Toilets	If there is a lock on the door there is no need to have separate male and female toilets. There should be enough for the site workforce. If there is no flushing water on site they must be chemical toilets.
Washing facilities	There should be a wash basin large enough to be able to wash up to the elbow. There should be soap, hot and cold water and, if you are working with dangerous substances, then showers are needed.
Drinking water	Clean drinking water should be available; either directly connected to the mains or bottled water. Employers must ensure that there is no contamination.
Dry room	This can operate also as a store room, which needs to be secure so that workers can leave their belongings there and also use it as a place to dry out if they have been working in wet weather, in which case a heater needs to be provided.
Work break area	This is a shelter out of the wind and rain, with a kettle, a microwave, tables and chairs. It should also have heating.

Table 1.7 Welfare facilities in the workplace

CASE STUDY

South Tyneside Homes

South Tyneside Council's Housing Company

Staying safe on site

Johnny McErlane finished his apprenticeship at South Tyneside Homes a year ago.

'I've been working on sheltered accommodation for the last year, so there are a lot of vulnerable and elderly people around. All the things I learnt at college from doing the health and safety exams comes into practice really, like taking care when using extension leads, wearing high-vis and correct footwear. It's not just about your health and safety, but looking out for others as well.

On the shelters, you can get a health and safety inspector who just comes around randomly, so you have to always be ready. It just becomes a habit once it's been drilled into you. You're health and safety conscious all the time.

The shelters also have a fire alarm drill every second Monday, so you've got to know the procedure involved there. When it comes to the more specialised skills, such as mouth-to-mouth and CPR, you might have a designated first aider on site who will have their skills refreshed regularly. Having a full first aid certificate would be valuable if you're working in construction.

You cover quite a bit of the first aid skills in college and you really have to know them because you're not always working on large sites. For example, you might be on the repairs team, working in people's houses where you wouldn't have a first aider, so you've got to have the basic knowledge yourself, just in case. All our vans have a basic first aid kit that's kept fully stocked.

The company keeps our knowledge current with these "toolbox talks", which are like refresher courses. They give you any new information that needs to be passed on to all the trades. It's a good way of keeping everyone up to date.'

Noise

Ear defenders are the best precaution to protect the ears from loud noises on site. Ear defenders are either basic ear plugs or ear muffs, which can be seen in Fig 1.13 on page 32.

The long-term impact of noise depends on the intensity and duration of the noise. Basically, the louder and longer the noise exposure, the more damage is caused. There are ways of dealing with this:

* Remove the source of the noise.

* Move the equipment away from those not directly working with it.

* Put the source of the noise into a soundproof area or cover it with soundproof material.

* Ask a supervisor if they can move all other employees away from that part of the site until the noise stops.

Substances hazardous to health

COSHH Regulations (see page 3) identify a wide variety of substances and materials that must be labelled in different ways.

Controlling the use of these substances is always difficult. Ideally, their use should be eliminated (stopped) or they should be replaced with something less harmful. Failing this, they should only be used in controlled or restricted areas. If none of this is possible then they should only be used in controlled situations.

If a hazardous situation occurs at work, then you should:

* ensure the area is made safe

* inform the supervisor, site manager, safety officer or other nominated person.

You will also need to report any potential hazards or near misses.

Personal hygiene

Construction sites can be dirty places to work. Some jobs will expose you to dust, chemicals or substances that can make contact with your skin or may stain your work clothing. It is good practice to wear suitable PPE as a first line of defence as chemicals can penetrate your skin. Whenever you have finished a job you should always wash your hands. This is certainly true before eating lunch or travelling home. It can be good practice to have dedicated work clothing, which should be washed regularly.

Always ensure you wash your hands and face and scrub your nails. This will prevent dirt, chemicals and other substances from contaminating your food and your home.

Make sure that you regularly wash your work clothing and either repair it or replace it if it becomes too worn or stained.

Health risks

The construction industry uses a wide variety of substances that could harm your health. You will also be carrying out work that could be a health risk to you, and you should always be aware that certain activities could cause long-term damage or even kill you if things go wrong. Unfortunately not all health risks are immediately obvious. It is important to make sure that from time to time you have health checks, particularly if you have been using hazardous substances. Table 1.8 outlines some potential health risks in a typical construction site.

Health risk	Potential future problems
Dust	The most dangerous potential dust is, of course, asbestos, which **should only be handled by specialists under controlled conditions**. But even brick dust and other fine particles can cause eye injuries, problems with breathing and even cancer.
Chemicals	Inhaling or swallowing dangerous chemicals could cause immediate, long-term damage to lungs and other internal organs. Skin problems include burns or skin can become very inflamed and sore. This is known as dermatitis.
Bacteria	Contact with waste water or soil could lead to a bacterial infection. The germs in the water or dirt could cause infection which will require treatment if they enter the body. The most extreme version is leptospirosis.
Heavy objects	Lifting heavy, bulky or awkward objects can lead to permanent back injuries that could require surgery. Heavy objects can also damage the muscles in all areas of the body.
Noise	Failure to wear ear defenders when you are exposed to loud noises can permanently affect your hearing. This could lead to deafness in the future.
Vibrating tools	Using machines that vibrate can cause a condition known as hand/arm vibration syndrome (HAVS) or vibration white finger, which is caused by injury to nerves and blood vessels. You will feel tingling that could lead to permanent numbness in the fingers and hands, as well as muscle weakness.
Cuts	Any open wound, no matter how small, leaves your body exposed to potential infections. Cuts should always be cleaned and covered, preferably with a waterproof dressing. The blood loss from deep cuts could make you feel faint and weak, which may be dangerous if you are working at height or operating machinery.
Sunlight	Most construction work involves working outside. There is a temptation to take advantage of hot weather and get a tan. But long-term exposure to sunshine means risking skin cancer so you should cover up and apply sun cream.
Head injuries	You should seek medical attention after any bump to the head. Severe head injuries could cause epilepsy, hearing problems, brain damage or death.

Table 1.8 Health risks in construction

HANDLING AND STORING MATERIALS AND EQUIPMENT

On a busy construction site it is often tempting not to even think about the potential dangers of handling equipment and materials. If something needs to be moved or collected you will just pick it up without any thought. It is also tempting just to drop your tools and other equipment when you have finished with them to deal with later. But abandoned equipment and tools can cause hazards both for you and for other people.

Safe lifting

Lifting or handling heavy or bulky items is a major cause of injuries on construction sites. So whenever you are dealing with a heavy load, it is important to carry out a basic risk assessment.

The first thing you need to do is to think about the job to be done and ask:

* Do I need to lift it manually or is there another way of getting the object to where I need it?

Consider any mechanical methods of transporting loads or picking up materials. If there really is no alternative, then ask yourself:

1. Do I need to bend or twist?
2. Does the object need to be lifted or put down from high up?
3. Does the object need to be carried a long way?
4. Does the object need to be pushed or pulled for a long distance?
5. Is the object likely to shift around while it is being moved?

If the answer to any of these questions is 'yes', you may need to adjust the way the task is done to make it safer.

Think about the object itself. Ask:

1. Is it just heavy or is it also bulky and an awkward shape?
2. How easy is it to get a good hand-hold on the object?
3. Is the object a single item or are there parts that might move around and shift the weight?
4. Is the object hot or does it have sharp edges?

Again, if you have answered 'yes' to any of these questions, then you need to take steps to address these issues.

It is also important to think about the working environment and where the lifting and carrying is taking place. Ask yourself:

1. Are the floors stable?

2. Are the surfaces slippery?

3. Will a lack of space restrict my movement?

4. Are there any steps or slopes?

5. What is the lighting like?

Before lifting and moving an object, think about the following:

* Check that your pathway is clear to where the load needs to be taken.

* Look at the product data sheet and assess the weight. If you think the object is too heavy or difficult to move then ask someone to help you. Alternatively, you may need to use a mechanical lifting device.

When you are ready to lift, gently raise the load. Take care to ensure the correct posture – you should have a straight back, with your elbows tucked in, your knees bent and your feet slightly apart.

Once you have picked up the load, move slowly towards your destination. When you get there, make sure that you do not drop the load but carefully place it down.

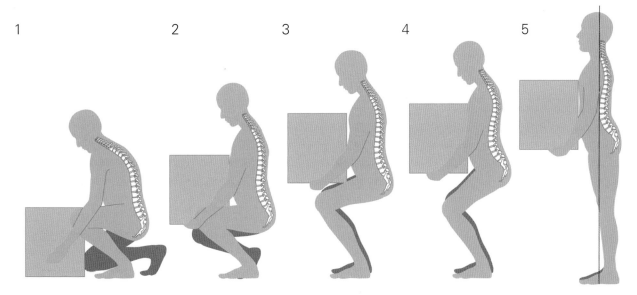

Figure 1.4 Take care to follow the correct procedure for lifting

Sack trolleys are useful for moving heavy and bulky items around. Gently slide the bottom of the sack trolley under the object and then raise the trolley to an angle of 45° before moving off. Make sure that the object is properly balanced and is not too big for the trolley.

Trailers and forklift trucks are often used on large construction sites, as are dump trucks. Never use these without proper training.

Figure 1.5 Pallet truck

Figure 1.6 Sack trolley

Site safety equipment

You should always read the construction site safety rules and when required wear your PPE. Simple things, such as wearing the right footwear for the right job, are important.

Safety equipment falls into two main categories:

* PPE – including hard hats, footwear, gloves, glasses and safety vests

* perimeter safety – this includes screens, netting and guards or clamps to prevent materials from falling or spreading.

Construction safety is also directed by signs, which will highlight potential hazards.

Safe handling of materials and equipment

All tools and equipment are potentially dangerous. It is up to you to make sure that they do not cause harm to yourself or others. You should always know how to use tools and equipment. This means either instruction from someone else who is experienced, or at least reading the manufacturer's instructions.

You should always make sure that you:

* use the right tool – don't be tempted to use a tool that is close to hand instead of the one that is right for the job

* wear your PPE – the one time you decide not to bother could be the time that you injure yourself

* never try to use a tool or a piece of equipment that you have not been trained to use.

You should always remember that if you are working on a building that was constructed before 2000 it may contain asbestos.

Correct storage

We have already seen that tools and equipment need to be treated with respect. Damaged tools and equipment are not only less effective at doing their job, they could also cause you to injure yourself.

Table 1.9 provides some pointers on how to store and handle different types of materials and equipment.

Materials and equipment	Safe storage and handling
Hand tools	Store hand tools with sharp edges either in a cover or a roll. They should be stored in bags or boxes. They should always be dried before putting them away as they will rust.
Power tools	Never carry them by the cable. Store them in their original carrying case. Always follow the manufacturer's instructions.
Wheelbarrows	Check the tyres and metal stays regularly. Always clean out after use and never overload.
Bricks and blocks	Never store more than two packs high. When cutting open a pack, be careful as the bricks could collapse.
Slabs and curbs	Store slabs flat on their edges on level ground, preferably with wood underneath to prevent damage. Store curbs the same way. To prevent weather damage, cover them with a sheet.
Tiles	Always cover them and protect them from damage as they are relatively fragile. Ideally store them in a hut or container.
Aggregates	Never store aggregates under trees as leaves will drop on them and contaminate them. Cover them with plastic sheets.
Plaster and plasterboard	Plaster needs to be kept dry, so even if stored inside you should take the precaution of putting the bags on pallets. To prevent moisture do not store against walls and do not pile higher than five bags. Plasterboard can be awkward to manage and move around. It also needs to be stored in a waterproof area. It should be stored flat and off the ground but should not be stored against walls as it may bend. Use a rotation system so that the materials are not stored in the same place for long periods.
Wood	Always keep wood in dry, well-ventilated conditions. If it needs to be stored outside it should be stored on bearers that may be on concrete. If wood gets wet and bends it is virtually useless. Always be careful when moving large cuts of wood or sheets of ply or MDF as they can easily become damaged.
Adhesives and paint	Always read the manufacturer's instructions. Ideally they should always be stored on clearly marked shelves. Make sure you rotate the stock using the older stock first. Always make sure that containers are tightly sealed. Storage areas must comply with fire regulations and display signs to advise of their contents.

Table 1.9 Safe storing and handling of materials and equipment

Waste control

The expectation within the building services industry is increasingly that working practices conserve energy and protect the environment. Everyone can play a part in this. For example, you can contribute by turning off hose pipes when you have finished using water, or not running electrical items when you don't need to.

Simple things, such as keeping construction sites neat and orderly, can go a long way to conserving energy and protecting the environment. A good way to remember this is Sort, Set, Shine, Standardise:

* Sort – sort and store items in your work area, eliminate clutter and manage deliveries.

* Set – everything should have its own place and be clearly marked and easy to access. In other words, be neat!

Figure 1.7 It's important to create as little waste as possible on the construction site

* Shine – clean your work area and you will be able to see potential problems far more easily.

* Standardise – by using standardised working practices you can keep organised, clean and safe.

Reducing waste is all about good working practice. By reducing wastage disposal, and recycling materials on site, you will benefit from savings on raw materials and lower transportation costs.

Planning ahead, and accurately measuring and cutting materials, means that you will be able to reduce wastage.

BASIC WORKING PLATFORMS AND ACCESS EQUIPMENT

Working at height should be eliminated or the work carried out using other methods where possible. However, there may be situations where you may need to work at height. These situations can include:

* roofing

* repair and maintenance above ground level

* working on high ceilings.

Any work at height must be carefully planned. Access equipment includes all types of ladder, scaffold and platform. You must always use a working platform that is safe. Sometimes a simple step ladder will be sufficient, but at other times you may have to use a tower scaffold.

Generally, ladders are fine for small, quick jobs of less than 30 minutes. However, for larger, longer jobs a more permanent piece of access equipment will be necessary.

Working platforms and access equipment: good practice and dangers of working at height

Table 1.10 outlines the common types of equipment used to allow you to work at heights, along with the basic safety checks necessary.

Equipment	Main features	Safety checks
Step ladder	Ideal for confined spaces. Four legs give stability	• Knee should remain below top of steps • Check hinges, cords or ropes • Position only to face work
Ladder	Ideal for basic access, short-term work. Made from aluminium, fibreglass or wood	• Check rungs, tie rods, repairs, and ropes and cords on stepladders • Ensure it is placed on firm, level ground • Angle should be no greater than 75° or 1 in 4
Mobile mini towers or scaffolds	These are usually aluminium and foldable, with lockable wheels	• Ensure the ground is even and the wheels are locked • Never move the platform while it has tools, equipment or people on it
Roof ladders and crawling boards	The roof ladder allows access while crawling boards provide a safe passage over tiles	• The ladder needs to be long enough and supported • Check boards are in good condition • Check the welds are intact • Ensure all clips function correctly
Mobile tower scaffolds	These larger versions of mini towers usually have edge protection	• Ensure the ground is even and the wheels are locked • Never move the platform while it has tools, equipment or people on it • Base width to height ratio should be no greater than 1:3
Fixed scaffolds and edge protection	Scaffolds fitted and sized to the specific job, with edge protection and guard rails	• There needs to be sufficient braces, guard rails and scaffold boards • The tubes should be level • There should be proper access using a ladder
Mobile elevated work platforms	Known as scissor lifts or cherry pickers	• Specialist training is required before use • Use guard rails and toe boards • Care needs to be taken to avoid overhead hazards such as cables

Table 1.10 Equipment for working at height and safety checks

You must be trained in the use of certain types of access equipment, like mobile scaffolds. Care needs to be taken when assembling and using access equipment. These are all examples of good practice:

* Step ladders should always rest firmly on the ground. Only use the top step if the ladder is part of a platform.

* Do not rest ladders against fragile surfaces, and always use both hands to climb. It is best if the ladder is steadied (footed) by someone at the foot of the ladder. Always maintain three points of contact – two feet and one hand.

* A roof ladder is positioned by turning it on its wheels and pushing it up the roof. It then hooks over the ridge tiles. Ensure that the access ladder to the roof is directly beside the roof ladder.

* A mobile scaffold is put together by slotting sections until the required height is reached. The working platform needs to have a suitable edge protection such as guard-rails and toe-boards. Always push from the bottom of the base and not from the top to move it, otherwise it may lean or topple over.

Figure 1.8 A tower scaffold

WORKING SAFELY WITH ELECTRICITY

It is essential whenever you work with electricity that you are competent and that you understand the common dangers. Electrical tools must be used in a safe manner on site. There are precautions that you can take to prevent possible injury, or even death.

Precautions

Whether you are using electrical tools or equipment on site, you should always remember the following:

* Use the right tool for the job.

* Use a transformer with equipment that runs on 110V.

* Keep the two voltages separate from each other. You should avoid using 230V where possible but, if you must, use a residual current device (RCD) if you have to use 230V.

* When using 110V, ensure that leads are yellow in colour.

* Check the plug is in good order.

* Confirm that the fuse is the correct rating for the equipment.

* Check the cable (including making sure that it does not present a tripping hazard).

* Find out where the mains switch is, in case you need to turn off the power in the event of an emergency.

* Never attempt to repair electrical equipment yourself.

* Disconnect from the mains power before making adjustments, such as changing a drill bit.

* Make sure that the electrical equipment has a sticker that displays a recent test date.

Visual inspection and testing is a three-stage process:

1. The user should check for potential danger signs, such as a frayed cable or cracked plug.

2. A formal visual inspection should then take place. If this is done correctly then most faults can be detected.

3. Combined inspections and **PAT** should take place at regular intervals by a competent person.

Watch out for the following causes of accidents – they would also fail a safety check:

KEY TERMS

PAT

– Portable Appliance Testing – regular testing is a health and safety requirement under the Electricity at Work Regulations (1989).

- damage to the power cable or plug
- taped joints on the cable
- wet or rusty tools and equipment
- weak external casing
- loose parts or screws
- signs of overheating
- the incorrect fuse
- lack of cord grip
- electrical wires attached to incorrect terminals
- bare wires.

When preparing to work on an electrical circuit, do not start until a permit to work has been issued by a supervisor or manager to a competent person.

Make sure the circuit is broken before you begin. A 'dead' circuit will not cause you, or anybody else, harm. These steps must be followed:

- Switch off – ensure the supply to the circuit is switched off by disconnecting the supply cables or using an isolating switch.
- Isolate – disconnect the power cables or use an isolating switch.
- Warn others – to avoid someone reconnecting the circuit, place warning signs at the isolation point.
- Lock off – this step physically prevents others from reconnecting the circuit.
- Testing – is carried out by electricians but you should be aware that it involves three parts:

  1. testing a voltmeter on a known good source (a live circuit) so you know it is working properly
  2. checking that the circuit to be worked on is dead
  3. rechecking your voltmeter on the known live source, to prove that it is still working properly.

It is important to make sure that the correct point of isolation is identified. Isolation can be next to a local isolation device, such as a plug or socket, or a circuit breaker or fuse.

The isolation should be locked off using a unique key or combination. This will prevent access to a main isolator until the work has been completed. Alternatively, the handle can be made detachable in the OFF position so that it can be physically removed once the circuit is switched off.

Dangers

You are likely to encounter a number of potential dangers when working with electricity on construction sites or in private houses. Table 1.11 outlines the most common dangers.

Danger	Identifying the danger
Faulty electrical equipment	Visually inspect for signs of damage. Equipment should be double insulated or incorporate an earth cable.
Damaged or worn cables	Check for signs of wear or damage regularly. This includes checking power tools and any wiring in the property.
Trailing cables	Cables lying on the ground, or worse, stretched too far, can present a tripping hazard. They could also be cut or damaged easily.
Cables and pipe work	Always treat services you find as though they are live. This is very important as services can be mistaken for one another. You may have been trained to use a cable and pipe locator that finds cables and metal pipes.
Buried or hidden cables	Make sure you have plans. Alternatively, use a cable and pipe locator, mark the positions, look out for signs of service connection cables or pipes and hand-dig trial holes to confirm positions.
Inadequate over-current protection	Check circuit breakers and fuses are the correct size current rating for the circuit. A qualified electrician may have to identify and label these.

Table 1.11 Common dangers when working with electricity

Each year there are around 1,000 accidents at work involving electric shocks or burns from electricity. If you are working in a construction site you are part of a group that is most at risk. Electrical accidents happen when you are working close to equipment that you think is disconnected but which is, in fact, live.

Another major danger is when electrical equipment is either misused or is faulty. Electricity can cause fires and contact with the live parts can give you an electric shock or burn you.

Different voltages

The two most common voltages that are used in the UK are 230V and 110V:

* 230V: this is the standard domestic voltage. But on construction sites it is considered to be unsafe and therefore 110V is commonly used.

* 110V: these plugs are marked with a yellow casement and they have a different shaped plug. A transformer is required to convert 230V to 110V.

Some larger homes, as well as industrial and commercial buildings, may have 415V supplies. This is the same voltage that is found on overhead electricity cables. In most houses and other buildings the voltage from these cables is reduced to 230V. This is what most electrical equipment works from. Some larger machinery actually needs 415V.

In these buildings the 415V comes into the building and then can either be used directly or it is reduced so that normal 230V appliances can be used.

Colour coded cables

Normally you will come across three differently coloured wires: Live, Neutral and Earth. These have standard colours that comply with European safety standards and to ensure that they are easily identifiable. However, in some older buildings the colours are different.

Wire type	Modern colour	Older colour
Live	Brown	Red
Neutral	Blue	Black
Earth	Yellow and Green	Yellow and Green

Table 1.12 Colour coding of cables

Working with equipment with different electrical voltages

You should always check that the electrical equipment that you are going to use is suitable for the available electrical supply. The equipment's power requirements are shown on its rating plate. The voltage from the supply needs to match the voltage that is required by the equipment.

Storing electrical equipment

Electrical equipment should be stored in dry and secure conditions. Electrical equipment should never get wet but – if it does happen – it should be dried before storage. You should always clean and adjust the equipment before connecting it to the electricity supply.

PERSONAL PROTECTIVE EQUIPMENT (PPE)

Personal protective equipment, or PPE, is a general term that is used to describe a variety of different types of clothing and equipment that aim to help protect against injuries or accidents. Some PPE you will use on a daily basis and others you may use from time to time. The type of PPE you wear depends on what you are doing and where you are. For example, the practical exercises in this book were photographed at a college, which has rules and requirements for PPE that are different to those on large construction sites. Follow your tutor's or employer's instructions at all times.

Types of PPE

PPE literally covers from head to foot. Here are the main PPE types.

Figure 1.9 A hi-vis jacket

Figure 1.10 Safety glasses and goggles

Figure 1.11 Hand protection

Figure 1.12 Head protection

Figure 1.13 Hearing protection

Protective clothing

Clothing protection such as overalls:

* provides some protection from spills, dust and irritants
* can help protect you from minor cuts and abrasions
* reduces wear to work clothing underneath.

Sometimes you may need waterproof or chemical-resistant overalls.

High visibility (hi-vis) clothing stands out against any background or in any weather conditions. It is important to wear high visibility clothing on a construction site to ensure that people can see you easily. In addition, workers should always try to wear light-coloured clothing underneath, as it is easier to see.

You need to keep your high visibility and protective clothing clean and in good condition.

Employers need to make sure that employees understand the reasons for wearing high visibility clothing and the consequences of not doing so.

Eye protection

For many jobs, it is essential to wear goggles or safety glasses to prevent small objects, such as dust, wood or metal, from getting into the eyes. As goggles tend to steam up, particularly if they are being worn with a mask, safety glasses can often be a good alternative.

Hand protection

Wearing gloves will help to prevent damage or injury to the hands or fingers. For example, general purpose gloves can prevent cuts, and rubber gloves can prevent skin irritation and inflammation, such as contact dermatitis caused by handling hazardous substances. There are many different types of gloves available, including specialist gloves for working with chemicals.

Head protection

Hard hats or safety helmets are compulsory on building sites. They can protect you from falling objects or banging your head. They need to fit well and they should be regularly inspected and checked for cracks. Worn straps mean that the helmet should be replaced, as a blow to the head can be fatal. Hard hats bear a date of manufacture and should be replaced after about 3 years.

Hearing protection

Ear defenders, such as ear protectors or plugs, aim to prevent damage to your hearing or hearing loss when you are working with loud tools or are involved in a very noisy job.

Respiratory protection

Breathing in fibre, dust or some gases could damage the lungs. Dust is a very common danger, so a dust mask, face mask or respirator may be necessary.

Make sure you have the right mask for the job. It needs to fit properly otherwise it will not give you sufficient protection.

Foot protection

Foot protection is compulsory on site, particularly if you are undertaking heavy work. Footwear should include steel toecaps (or equivalent) to protect feet against dropped objects, midsole protection (usually a steel plate) to protect against puncture or penetration from things like nails on the floor and soles with good grip to help prevent slips on wet surfaces.

Figure 1.14 Respiratory protection

Legislation covering PPE

The most important piece of legislation is the Personal Protective Equipment at Work Regulations (1992). It covers all sorts of PPE and sets out your responsibilities and those of the employer. Linked to this are the Control of Substances Hazardous to Health (2002) and the Provision and Use of Work Equipment Regulations (1992 and 1998).

Storing and maintaining PPE

All forms of PPE will be less effective if they are not properly maintained. This may mean examining the PPE and either replacing or cleaning it, or if relevant testing or repairing it. PPE needs to be stored properly so that it is not damaged, contaminated or lost. Each type of PPE should have a CE mark. This shows that it has met the necessary safety requirements.

Importance of PPE

PPE needs to be suitable for its intended use and it needs to be used in the correct way. As a worker or an employee you need to:

* make sure you are trained to use PPE

* follow your employer's instructions when using the PPE and always wear it when you are told to do so

* look after the PPE and if there is a problem with it report it.

Your employer will:

* know the risks that the PPE will either reduce or avoid

* know how the PPE should be maintained

* know its limitations.

Consequences of not using PPE

The consequences of not using PPE can be immediate or long-term. Immediate problems are more obvious, as you may injure yourself. The longer-term consequences could be ill health in the future. If your employer has provided PPE, you have a legal responsibility to wear it.

FIRE AND EMERGENCY PROCEDURES

If there is a fire or an emergency, it is vital that you raise the alarm quickly. You should leave the building or site and then head for the **assembly point.**

When there is an emergency a general alarm should sound. If you are working on a larger and more complex construction site, evacuation may begin by evacuating the area closest to the emergency. Areas will then be evacuated one-by-one to avoid congestion of the escape routes.

Figure 1.15 Assembly point sign

Three elements essential to creating a fire

Three ingredients are needed to make something combust (burn):

* oxygen * heat * fuel.

The fuel can be anything which burns, such as wood, paper or flammable liquids or gases, and oxygen is in the air around us, so all that is needed is sufficient heat to start a fire.

The fire triangle represents these three elements visually. By removing one of the three elements the fire can be prevented or extinguished.

Figure 1.16 The fire triangle

How fire is spread

Fire can easily move from one area to another by finding more fuel. You need to consider this when you are storing or using materials on site, and be aware that untidiness can be a fire risk. For example, if there are wood shavings on the ground the fire can move across them, burning up the shavings.

Heat can also transfer from one source of fuel to another. If a piece of wood is on fire and is against or close to another piece of wood, that too will catch fire and the fire will have spread.

On site, fires are classified according to the type of material that is on fire. This will determine the type of fire-fighting equipment you will need to use. The five different types of fire are shown in Table 1.13.

Class of fire	Fuel or material on fire
A	Wood, paper and textiles
B	Petrol, oil and other flammable liquids
C	LPG, propane and other flammable gases
D	Metals and metal powder
E	Electrical equipment

Table 1.13 Different classes of fire

There is also F, cooking oil, but this is less likely to be found on site, except in a kitchen.

Taking action if you discover a fire and fire evacuation procedures

During induction, you will have been shown what to do in the event of a fire and told about assembly points. These are marked by signs and somewhere on the site there will be a map showing their location.

If you discover a fire you should:

* sound the alarm

* not attempt to fight the fire unless you have had fire marshal training

* otherwise stop work, do not collect your belongings, do not run, and do not re-enter the site until the all clear has been given.

Different types of fire extinguishers

Extinguishers can be effective when tackling small localised fires. However, you must use the correct type of extinguisher. For example, putting water on an oil fire could make it explode. For this reason, you should not attempt to use a fire extinguisher unless you have had proper training.

When using an extinguisher it is important to remember the following safety points:

* Only use an extinguisher at the early stages of a fire, when it is small.

* The instructions for use appear on the extinguisher.

* If you do choose to fight the fire because it is small enough, and you are sure you know what is burning, position yourself between the fire and the exit, so that if it doesn't work you can still get out.

Type of fire risk	Fire class Symbol	White label Water	Cream label Foam	Black label Carbon dioxide	Blue label Dry powder	Yellow label Wet chemical
A – Solid (e.g. wood or paper)	A	✓	✓	✗	✓	✓
B – Liquid (e.g. petrol)	B	✗	✓	✓	✓	✗
C – Gas (e.g. propane)	C	✗	✗	✓	✓	✗
D – Metal (e.g. aluminium)	D METAL	✗	✗	✗	✓	✗
E – Electrical (i.e. any electrical equipment)	E	✗	✗	✓	✓	✗
F – Cooking oil (e.g. a chip pan)	F	✗	✗	✗	✗	✓

Table 1.14 Types of fire extinguishers

There are some differences you should be aware of when using different types of extinguisher:

- CO_2 extinguishers – do not touch the nozzle; simply operate by holding the handle. This is because the nozzle gets extremely cold when ejecting the CO_2, as does the canister. Fires put out with a CO_2 extinguisher may reignite, and you will need to ventilate the room after use.

- Powder extinguishers – these can be used on lots of kinds of fire, but can seriously reduce visibility by throwing powder into the air as well as on the fire.

SIGNS AND SAFETY NOTICES

In a well-organised working environment safety signs will warn you of potential dangers and tell you what to do to stay safe. They are used to warn you of hazards. Their purpose is to prevent accidents. Some will tell you what to do (or not to do) in particular parts of the site and some will show you where things are, such as the location of a first aid box or a fire exit.

Types of signs and safety notices

There are five basic types of safety sign, as well as signs that are a combination of two or more of these types. These are shown in Table 1.15.

Type of safety sign	What it tells you	What it looks like	Example
Prohibition sign	Tells you what you must *not* do	Usually round, in red and white	Do not use ladder
Hazard sign	Warns you about hazards	Triangular, in yellow and black	Caution Slippery floor
Mandatory sign	Tells you what you *must* do	Round, usually blue and white	Masks must be worn in this area
Safe condition or information sign	Gives important information, e.g. about where to find fire exits, assembly points or first aid kit, or about safe working practices	Green and white	First aid
Firefighting sign	Gives information about extinguishers, hydrants, hoses and fire alarm call points, etc.	Red with white lettering	Fire alarm call point
Combination sign	These have two or more of the elements of the other types of sign, e.g. hazard, prohibition and mandatory		DANGER Isolate before removing cover

Table 1.15 Different types of safety signs

TEST YOURSELF

1. Which of the following requires you to tell the HSE about any injuries or diseases?

   a. HASAWA

   b. COSHH

   c. RIDDOR

   d. PUWER

2. What is a prohibition notice?

   a. An instruction from the HSE to stop all work until a problem is dealt with

   b. A manufacturer's announcement to stop all work using faulty equipment

   c. A site contractor's decision not to use particular materials

   d. A local authority banning the use of a particular type of brick

3. Which of the following is considered a major injury?

   a. Bruising on the knee

   b. Cut

   c. Concussion

   d. Exposure to fumes

4. If there is an accident on a site who is likely to be the first to respond?

   a. First aider

   b. Police

   c. Paramedics

   d. HSE

5. Which of the following is a summary of risk assessments and is used for high risk activities?

   a. Site notice board

   b. Hazard book

   c. Monitoring statement

   d. Method statement

6. Some substances are combustible. Which of the following are examples of combustible materials?

   a. Adhesives

   b. Paints

   c. Cleaning agents

   d. All of these

7. What is dermatitis?

   a. Inflammation of the skin

   b. Inflammation of the ear

   c. Inflammation of the eye

   d. Inflammation of the nose

8. Screens, netting and guards on a site are all examples of which of the following?

   a. PPE

   b. Signs

   c. Perimeter safety

   d. Electrical equipment

9. Which of the following are also known as scissor lifts or cherry pickers?

   a. Bench saws

   b. Hand-held power tools

   c. Cement additives

   d. Mobile elevated work platforms

10. In older properties the neutral electricity wire is which colour?

    a. Black

    b. Red

    c. Blue

    d. Brown

Chapter 2

Unit CSA–L2Core04

UNDERSTAND INFORMATION, QUANTITIES AND COMMUNICATION WITH OTHERS

LEARNING OUTCOMES

LO1: Know how to interpret and produce information relating to construction

LO2: Understand how to estimate quantities of resources

LO3: Understand how to communicate workplace requirements efficiently

INTRODUCTION

The aim of this chapter is to:

* help you interpret and produce information relating to construction

* show you how to estimate quantities of resources

* enable you to communicate workplace requirements effectively to all levels of the construction team.

INTERPRETING AND PRODUCING INFORMATION

Even quite simple construction projects will require documents. These provide you with the necessary information that you will need to do the job. The documents are produced by a range of different people and each document has a different purpose. Together they give you the full picture of the job, from the basic outline through to the technical specifications.

Types of supporting information

Supporting information can be found in a variety of different types of documents. These include:

* drawings and plans

* programmes of work

* procedures

* specifications

* policies

* schedules

* manufacturers' technical information

* organisational documentation

* training and development records

* risk and method statements

* Construction (Design and Management) (CDM) Regulations

* Building Regulations.

Drawings and plans

Drawings are an important part of construction work. You will need to understand how drawings provide you with the information required to carry out the work. The drawings show what the building will look like and how it will be constructed. This means that there are several different drawings of the building from different viewpoints. In practice, most of the drawings are shown on the same sheet.

Block plans

Block plans show the construction site and the surrounding area. Normally block plans are at a ratio of 1:2500 and 1:1250. This means that 1 millimetre on a block plan is equal to 2,500 mm (2.5 m) or 1,250 mm (1.25 m) on the ground.

Site plan

Location drawings are sometimes known as site plans. The site plan drawing shows what is basically planned for the site. It is an important drawing because it has been created in order to get approval for the project from planning committees or funding sources. In most cases the site plan is actually an architectural plan, showing the basic arrangement of buildings and any landscaping.

The site plan will usually show:

* directional orientation (i.e. the north point)

* location and size of the building or buildings

* existing structures

* clear measurements

* colours and materials to be used.

General location

Location drawings show the site or building in relation to its surroundings. It will therefore show details such as boundaries, other buildings and roads. It will also contain other vital information, including:

* access

* drainage

* sewers

* the north point.

The drawing will have a title and will show the scale. A job or project number will help to identify it easily, and it will also have an address, the date when the drawing was done and the name of the client. A version number will also be on the drawing, with an amendment date if there have been any changes. It is important to make sure you have the latest drawing.

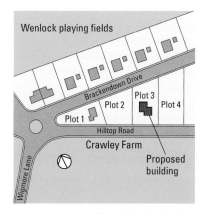

Figure 2.1 Block plan

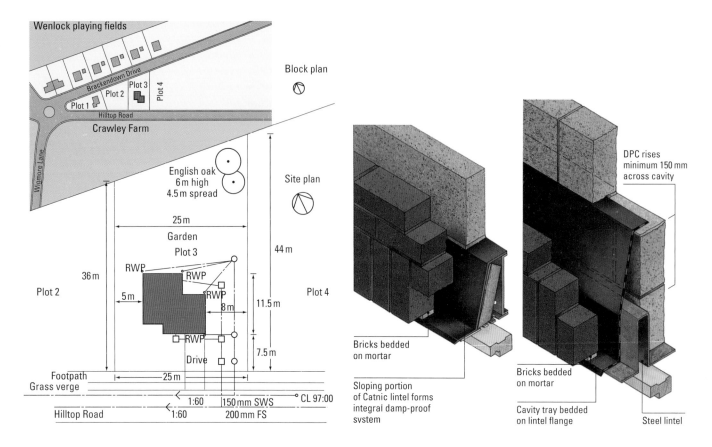

Figure 2.2 Location plan

Figure 2.3 Assembly drawing

Normally location drawings are either 1:500 or 1:200 (that is, 1 mm of the drawing represents 500 mm or 200 mm on the ground).

Assembly

These are detailed drawings that illustrate the different elements and components of the construction. They are likely to be 1:20, 1:10 or 1:5 (1 mm of the drawing represents 20 mm, 10 mm or 5 mm on the ground). This larger scale allows more detail to be shown, to ensure accurate construction.

Sectional

These drawings aim to provide:

* vertical dimensions

* constructional details

* horizontal sections.

They can be used to show the height of ground levels, damp-proof courses, foundations and other aspects of the construction.

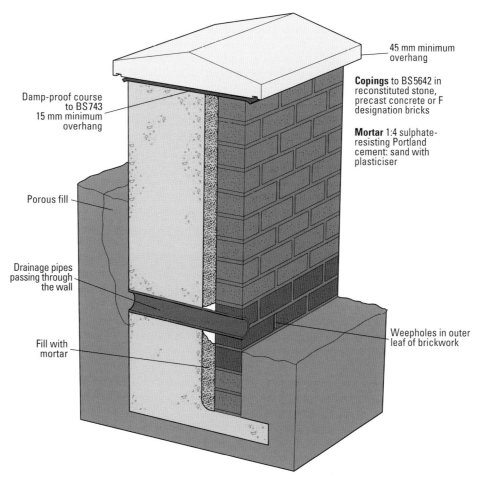

45 mm minimum overhang

Damp-proof course to BS 743 15 mm minimum overhang

Copings to BS 5642 in reconstituted stone, precast concrete or F designation bricks

Mortar 1:4 sulphate-resisting Portland cement: sand with plasticiser

Porous fill

Drainage pipes passing through the wall

Fill with mortar

Weepholes in outer leaf of brickwork

Figure 2.4 Section drawing of an earth retaining wall

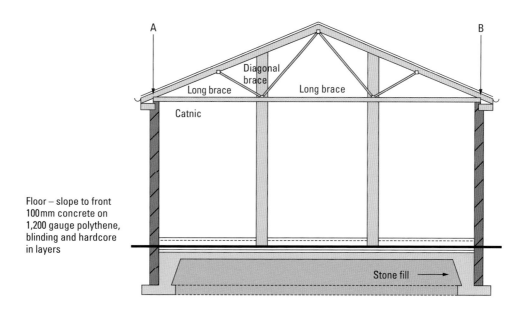

A

B

Diagonal brace

Long brace

Long brace

Catnic

Floor – slope to front 100mm concrete on 1,200 gauge polythene, blinding and hardcore in layers

Stone fill

Figure 2.5 Section drawing of a garage

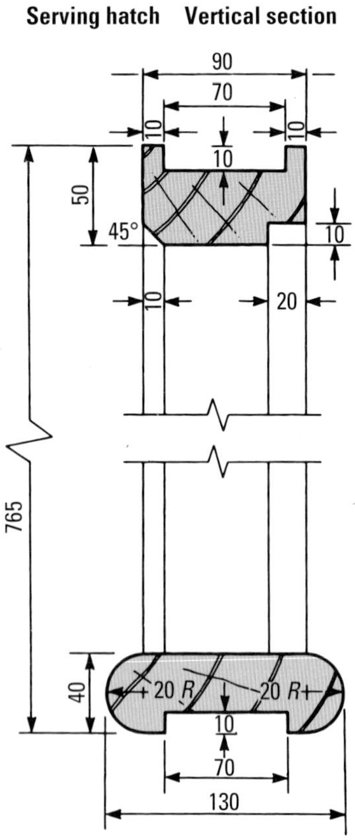

Serving hatch Vertical section
90
70
10
10
10
50
45°
10
10
20
765
40
20 R
20 R
10
70
130

Figure 2.6 Detail drawing

Details

These drawings show how a component needs to be manufactured. They can be shown in various scales, but mainly 1:10, 1:5 and 1:1 (the same size as the actual component if it is small).

Orthographic projection (first angle)

First angle projection is a view that represents the side of the object as if you were standing away from it, as can be seen in Fig 2.7.

Isometric projection

Isometric projection is a way of representing three-dimensional objects in two dimensions, as can also be seen in Fig 2.7. All horizontal lines are drawn at 30°.

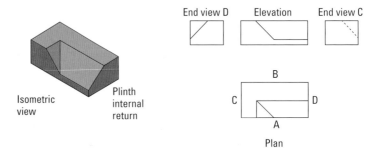

Isometric view
Plinth internal return
End view D
Elevation
End view C
B
C
D
A
Plan

Figure 2.7 First angle projection

Programmes of work

Programmes of work show the actual sequence of any work activities on a construction project. Part of the work programme plan is to show target times. They are usually shown in the form of a bar or Gantt chart (a special kind of bar chart), as can be seen in Fig 2.8.

In this figure:

- on the left hand side all of the tasks are listed – note this is in logical order
- on the right the blocks show the target start and end date for each of the individual tasks
- the timescale can be days, weeks or months.

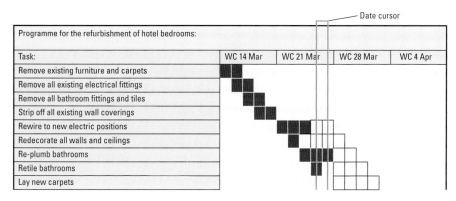

Figure 2.8 Single line contract plan Gantt chart

Far more complex forms of work programmes can also be created. The Gantt chart shown below (Fig 2.9) shows the construction of a house.

This is a more complex example of a bar chart:

* There are two lines – they show the target dates and actual dates. The actual dates are shaded, showing when the work actually began and how long it actually took.

* If this bar chart is kept up to date an accurate picture of progress and estimated completion time can be seen.

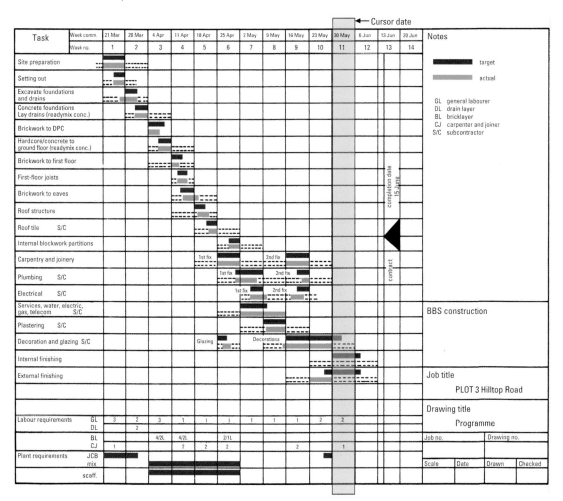

Figure 2.9 Gantt chart for the construction of a house

Procedures

When you work for a construction company it will have a series of procedures which you will have to follow. A good example is the emergency procedure. This will explain precisely what is required in the case of an emergency on site and who will have responsibility for carrying out particular duties. Procedures are there to show you the right way of doing something.

Another good example of a procedure is the procurement or buying procedure. This will outline:

* who is authorised to buy what, and how much individuals are allowed to spend

* any forms or documents that have to be completed when buying.

Specifications

In addition to drawings it is usually necessary to have documents known as specifications. These provide much more information, as can be seen in Fig 2.10.

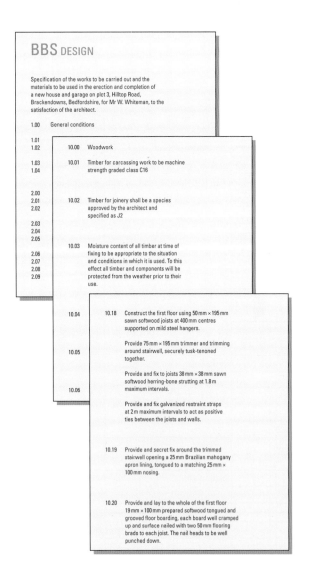

The specifications give you a precise description. They will include:

* the address and description of the site

* on-site services (e.g. water and electricity)

* materials description, outlining the size, finish, quality and tolerances

* specific requirements, such as the individual who will authorise or approve work carried out

* any restrictions on site, such as working hours.

Policies

Policies are sets of principles or a programme of actions. These are two good examples:

* Environmental policy – how the business goes about protecting the environment.

* Safety policy – how the business deals with health and safety matters and who is responsible for monitoring and maintaining it.

You will normally find both policies and procedures in site rules. These are usually explained to each new employee when they first join the company. Sometimes there may be additional site rules, depending on the job and the location of the work.

Figure 2.10 Extracts from a typical specification

Schedules

Schedules are cross-referenced to drawings that have been prepared by an architect. They will show specific design information. Usually they are prepared for jobs that will be carried out regularly on site, such as:

* working on windows, doors, floors, walls or ceilings

* working on drainage, lintels or sanitary ware.

A schedule can be seen in Fig 2.11.

The schedule is very useful for a number of purposes, such as:

* working out the quantities of materials needed

* ordering materials and components and then checking them against deliveries

* locating where specific materials will be used.

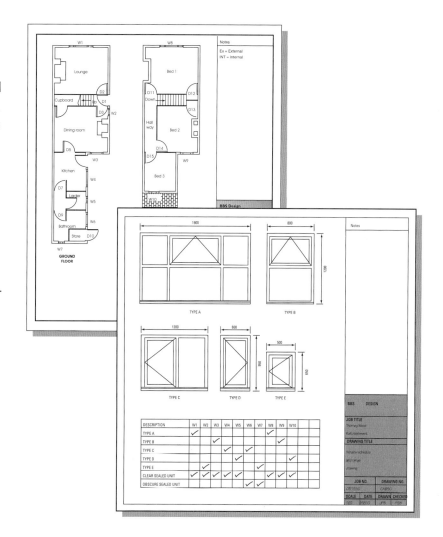

Figure 2.11 Typical windows schedule, range drawing and floor plans

Manufacturers' technical information

Almost everything that is bought to be used on site will come with a variety of types of information. The basic technical information provided will show what the equipment or material is intended to be used for, how it should be stored and any particular requirements it may have, such as for handling or maintenance.

Technical information from the manufacturer can come from a variety of different sources. These may include:

* printed or downloadable data sheets

* printed or downloadable user instructions

* manufacturers' catalogues or brochures

* manufacturers' websites.

Organisational documentation

The potential list of organisational documentation and paperwork is extensive. These are outlined in Table 2.1. Examples can be seen in Figs 2.12 to 2.16.

Document	Purpose
Timesheet	Record of hours that you have worked and the jobs that you have carried out. This is used to help work out your wages and the total cost of the job.
Day worksheet	This details work that has been carried out without providing an estimate beforehand. It usually includes repairs or extra work and alterations.
Variation order	Provided by the architect and given to the builder, showing any alterations, additions or omissions to the original job.
Confirmation notice	Provided by the architect to confirm any verbal instructions.
Daily report or site diary	This covers things that might affect the project like detailed weather conditions, late deliveries or site visitors.
Orders and requisitions	These are order forms, requesting the delivery of materials.
Delivery notes	These are provided by the supplier of materials as a list of all materials being delivered. These need to be checked against materials actually delivered. The buyer will sign the delivery note when they are happy with the delivery.
Delivery records	These are lists of all materials that have been delivered on site.
Memorandum	These are used for internal communications and are usually brief.
Letters	These are used for external communications, usually to customers or suppliers.
Fax	Even though email is commonly used, the industry still uses faxes, as they provide an exact copy of an original document.

Table 2.1 Types of organisational documentation

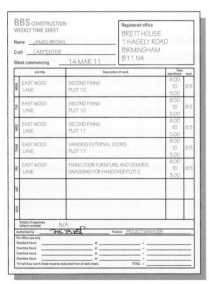

Figure 2.12 Timesheet

Figure 2.13 Day worksheet

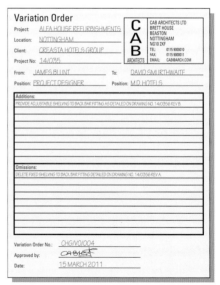

Figure 2.14 Variation order

Training and development records

Training and development is an important part of any job, as it ensures that employees have all the skills and knowledge that they need to do their work. Most medium to large employers will have training policies that set out how they intend to do this.

Employers will have a range of different documents to keep records and to make sure that they are on track. These documents will record all the training that an employee has undertaken.

Training can take place in a number of different ways and different places. It can include:

* induction

* toolbox talks

* in-house training

* specialist training

* training or education leading to formal qualifications.

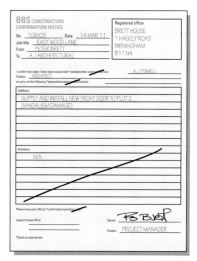

Figure 2.15 Confirmation notice

Checking information for conformity

The information to be checked can include drawings, programmes of work, schedules, policies, procedures, specifications and so on. The term 'conformity' in this sense means:

* making sure that any part of the assembly or component is suitable for the job

* making sure that the standard of work meets the necessary performance requirements.

This may mean that there could be an industry or trade standard that will need to be followed. The actual job or client may also require specific standards.

Figure 2.16 Daily report or site diary

Interpreting construction specifications

It would be difficult to put in all of the details in full, so symbols, hatchings and abbreviations are used to simplify the drawings. All of these symbols or hatchings are drawn to follow BS1192. The symbols cover various types of brickwork and blockwork, as well as concrete, hard core and insulation, as can be seen in Fig 2.17.

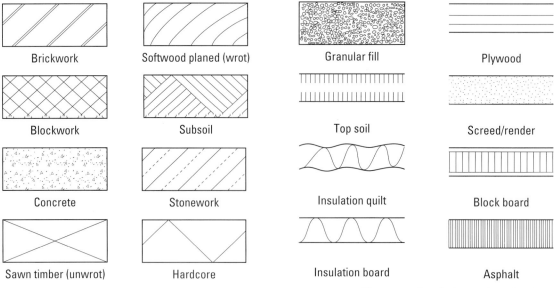

Figure 2.17 Symbols used on drawings

Abbreviation	Meaning
bwk	Refers to all types of brickwork
conc	Refers to areas that will be concreted
dpc	Refers to all types of damp-proof course
fdn	Refers to foundations that are required
insul	Refers to location and type of insulation
rwg	Refers to location of rainwater gulleys
svp	Refers to location and type of soil and vent pipe

Table 2.2 Abbreviations used in construction drawings

Common abbreviations

As we have already seen, covering the drawing with full detail would make it hard to read, so abbreviations are used. Table 2.2 outlines some examples of abbreviations that you will need to become familiar with.

Drawing equipment and its uses

Some basic equipment is necessary in order to produce drawings. These items are outlined in Table 2.3.

Equipment	Explanation and use
Scale rule	This is an essential piece of equipment. It needs to have 1:5/1:50, 1:10/1:100, 1:20/1:200 and 1:250/1:2500.
Set square	You will need to have a pair of these, or an adjustable square. If it is adjustable then you need to be able to create angles of up to 90°. The set square on the shortest side should be at least 150 mm. You will need the ability to create 30, 45, 60 and 90° angles.
Protractor	A protractor is essential to be able to measure angles up to and including 180°.
Compass	Compasses are used to create circles or arcs. It is also advisable to have a divider so that you can easily transfer measurements and dividing lines.
Pencils	For drawings you will need a 2H, 3H or 4H pencil. For sketching and darkening outlines you will need an HB pencil. You will need to keep these sharp.

Table 2.3 Drawing equipment required

In addition to this you will also need at least an A2 size drawing board that has a parallel rule (these may be provided by your college). It is also useful to have an eraser.

Scales used to produce construction drawings

When the plans for individual buildings or construction sites are drawn up they have to be scaled down so that they will fit on a manageable size of paper. It is important to remember that drawings are not sketches and that they are drawn to scale. This means that they are:

* exact and accurate

* in proportion to the real construction.

You can work out the dimensions by using the scale rule when measuring the drawings. There are several common scales used and the measurement is usually metric:

* 1:2500 – the drawing is 2,500 times smaller than the real object

* 1:100 – the drawing is 100 times smaller than the real object

* 1:50 – the drawing is 50 times smaller than the real object

* 1:20 – the drawing is 20 times smaller than the real object

* 1:10 – the drawing is 10 times smaller than the real object

* 1:5 – the drawing is 5 times smaller than the real object

* 1:2 – the drawing is 2 times smaller than the real object.

ESTIMATING QUANTITIES OF RESOURCES

Working out the quantity and cost of resources that are needed to do a particular job is, perhaps, one of the most difficult tasks. In most cases you or the company you work for will be asked to provide a price for the work.

It is generally accepted that there are three ways of doing this:

* estimate – an approximate calculation based on available information

* quotation – which is a fixed price

* tender – tendering is a process of allowing various parties to price for the same work. The process can be open or closed. This usually means that the result is fair.

As we will see a little later in this section, these three ways of costing are very different and each of them has its own problems.

Methods used to estimate quantities

Obviously past experience will help you to quickly estimate the amount of materials that will be needed on particular construction projects. This is also true of working out the best place to buy materials and how much the labour costs will be to get the job finished.

Many businesses will use the *Hutchins UK Building Costs Blackbook,* which provides a construction cost guide. It breaks down all types of work and shows an average cost for each of them.

Computerised estimating packages are available, which will give a comprehensive detailed estimate that looks very professional. This will also help to estimate quantities and timescales.

The alternative is of course to carry out a numerical calculation. It is therefore important to have the right resources upon which to base these calculations. These could be working drawings, schedules or other documents.

Usually all this involves making additions, subtractions, multiplications and divisions. In order to work out the amount of materials you will need for a construction project you will need to know some basic information:

* What does the job entail? How complex is it, and how much labour is required?

* What materials will be used?

* What are the costs of the materials?

Measurement

The standard unit for measurement is the metre (m). There are 100 centimetres (cm) and 1,000 millimetres (mm) in a metre. It is important to remember that drawings and plans have different scales, so these need to be converted to work out the quantities of materials required.

The most basic thing to work out is length (see Fig 2.18), from which you can calculate perimeter, area and then volume, capacity, mass and weight, as can be seen in Table 2.4.

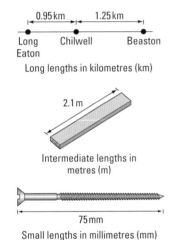

Figure 2.18 Length in metres and millimetres

Measurement	Explanation
Length	This is the distance from one end to the other. This could be measured in metres or millimetres, depending on the job.
Perimeter	This helps you work out the distance around a shape, such as the size of a room or a garden. It will help you estimate the length of a wall, for example. You just need to measure each side and then add them together (see Fig 2.20).
Area	You can work out the area of a room, for example, by measuring the length and the width of the room. Then you multiply the width by the length to give the number of square metres (m²) (see Fig 2.20).
Volume and capacity	Volume shows how much space is taken up by an object, such as a room. Again this is simply worked out by multiplying the width of the room by its length and then by its height. This gives you the number of cubic metres (m³). Capacity works in exactly the same way but instead of showing the figure as cubic metres you show it as litres. This is ideal if you are trying to work out the capacity of a water tank or a garden pond (see Fig 2.19).
Mass or weight	Mass is measured usually in kilograms or in grammes. Mass is the actual weight of a particular object, such as a brick.

Table 2.4 Working out measurements

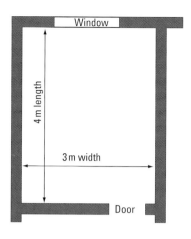

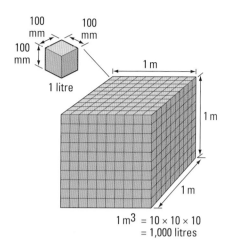

Figure 2.19 Measuring area and perimeter

Figure 2.20 Relationship between volume and capacity

Formulae

These can appear to be complicated, but using formulae is essential for working out quantities of materials. Each of the formulae is related to different shapes. In construction work you will often have to work out quantities of materials needed for odd shaped areas.

Area

To work out the area of a triangular shape, you use the following formula:

$$\text{Area (A)} = \text{Base (B)} \times \frac{\text{Height (H)}}{2}$$

So if a triangle has a base of 4.5 and a height of 3.5 the calculation is:

$$4.5 \times \frac{3.5}{2}$$

$$\text{Or } 4.5 \times 3.5 = \frac{15.75}{2} = 7.875\,\text{m}^2$$

Height

If you want to work out the height of a triangle you switch the formulae around. To give:

$$\text{Height} = 2 \times \frac{\text{Area}}{\text{Base}}$$

Perimeter

To work out the perimeter of a rectangle you use the formula:

$$\text{Perimeter} = 2 \times (\text{Length} + \text{Width})$$

It is important to remember this because you need to count the length and the width twice to ensure you have calculated the total distance around the object.

Circles

To work out the circumference or perimeter of a circle you use the formula:

$$\text{Circumference} = \pi\ (\text{pi}) \times \text{Diameter}$$

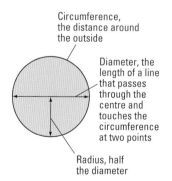

Circumference, the distance around the outside

Diameter, the length of a line that passes through the centre and touches the circumference at two points

Radius, half the diameter

Figure 2.22 Parts of a circle

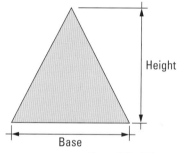

Height

Base

Figure 2.21 Triangle

π (pi) is always the same for all circles and is 3.142.

Diameter is the length of the widest part.

If you know the circumference and need to work out the diameter of the circle the formula is:

$$\text{Diameter} = \frac{\text{Circumference}}{\pi\ (\text{pi})}$$

For example if a circle has a circumference of 15.39 m then to work out the diameter:

$$\frac{15.39}{3.142} = 4.89\,\text{m}$$

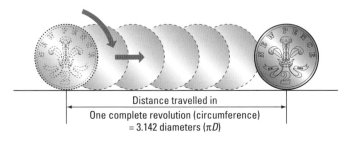

Distance travelled in
One complete revolution (circumference)
= 3.142 diameters (πD)

Figure 2.23 Relationship between circumference and diameter

Complex areas

Land, for example, is rarely square or rectangular. It is made up of odd shapes. You should never feel overwhelmed by complex areas, as all you need to do is to break them down into regular shapes.

By accurately measuring the perimeter you can then break down the shape into a series of triangles or rectangles. All that you need to do then is to work out the area of each of the shapes within the overall shape and add them up together.

Shape		Area equals	Perimeter equals
Square		AA (or A multiplied by A)	4A (or A multiplied by 4)
Rectangle		LB (or L multiplied by B)	2(L+B) (or L plus B multiplied by 2)

Shape	Area equals	Perimeter equals
Trapezium	$\frac{(A+B)H}{2}$ (or A plus B multiplied by H and then divided by 2)	A+B+C+D
Triangle	$\frac{BH}{2}$ (or B multiplied by H and then divided by 2)	A+B+C
Circle	πr^2 (or r multiplied by itself and then multiplied by pi (3.142))	πd or $2\pi r$

Table 2.5 Calculating complex areas

Volume

Sometimes it is necessary to work out the volume of an object, such as a cylinder or the amount of concrete needed. All that needs to be done is to work out the base area and then multiply that by the height.

For a concrete volume, if a 1.2 m square needs 3 m of height then the calculation is:

$$1.2 \times 1.2 \times 3 = 4.32\,m^3$$

To work out the volume of a cylinder you need to know the base area × the height. The formula is:

$$\pi r^2 \times H$$

So if a cylinder has a radius (r) of 0.8 and a height of 3.5 m then the calculation is:

$$3.142 \times 0.8 \times 0.8 \times 3.5 = 7.038\,m^3$$

Pythagoras

Pythagoras' theorem is used to work out the length of the sides of right-angled triangles. It states that:

In all right-angled triangles the square of the longest side is equal to the sum of the squares of the other two sides (that is, the length of a side multiplied by itself).

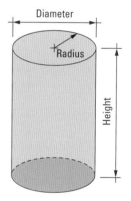

Figure 2.24 Cylinder

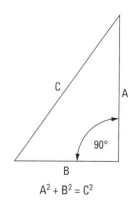

$A^2 + B^2 = C^2$

Figure 2.25 Pythagoras' theorem

Measuring materials

Using simple measurements and formulae can help you work out the amount of materials you will need. This is all summarised in Table 2.6.

Material	Measurement
Timber	To work out the linear run of a cubic metre of timber of a given cross sectional area, divide a square metre by the cross sectional area of one piece.
Flooring	To work out the amount of flooring for a particular area in metres^2 multiply the width of the floor by the length of the floor.
Stud walling, rafters and joists	Measure the distance that the stud partition will cover then divide that distance by a specified spacing and add 1. This will give you the number of spaces between each stud.
Fascias, barges and soffits	Measure the length and then add 10% for waste; however, this will depend on the nearest standard metric size of timber available.
Skirting, dado, picture rails and coving	You need to work out the perimeter of the room and then subtract any doorways or other openings. Again, add 10% for waste.
Bricks and mortar	Half brick walls use 60 bricks per metre squared and one brick walls use double that amount. You should add 5 per cent to take into account any cutting or damage. For mortar assume that you will need 1 kg for each brick.

Table 2.6 Working out materials required

How to cost materials

Once you have found out the quantity of materials necessary, you need to find out the price of those materials. You then do the costing by simply multiplying those prices by the amount of materials actually needed.

CASE STUDY

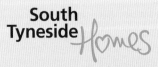

South Tyneside

South Tyneside Council's Housing Company

It's important to get it right

Glen Campbell is a team leader at South Tyneside Homes.

'Your English and maths skills really are important. As an apprentice, you have to be able to communicate properly – to get information and materials back and forth between tradespeople and yourself, to be able to sit and put a little drawing down, to label things up, and to take information off drawings – especially on the capital works jobs. You're reading and writing stuff down all the time... even your timesheets because they have to be accurate.

When it comes to your maths skills, you're using measurement all the time. If you get measurements wrong, you're not making the money. For example, if you're using the wrong size timber for a roof – the drawing says you've got to use 200 × 50 mm joists and then you go and use ones that are 150 mm – it's either going to cost you more to go back and get it right, or it's not going to be able to take that stress load once the roof goes on. In the end it could even collapse.'

Materials and purchasing systems

Many builders and companies will have preferred suppliers of materials. Many of them will already have negotiated discounts based on their likely spending with that supplier over the course of a year. The supplier will then be organised to supply them at an agreed price.

In other cases, builders may shop around to find the best price for the materials that match the specification. The lowest price may not necessarily be the best one to go for. All materials need to be of a sufficient quality. The other key consideration is whether the materials are immediately available for delivery.

It is vital that suppliers are reliable and that they have sufficient materials in stock. Delays in deliveries can cause major setbacks on site. It is not always possible to warn suppliers that materials will be needed, but a well-run site should be able to anticipate the materials that are needed and put in the orders within good time.

Large quantities may be delivered direct from the manufacturer straight to site. This is preferable when dealing with items where colours must be consistent.

Comparing estimated labour rates

The cost of labour for particular jobs is based on the hourly charge-out rate for that individual or group of individuals multiplied by the time it would take to complete the job.

Labour rates can depend on:

* the expertise of the construction worker
* the size of the business they work for
* the part of the country in which the work is being carried out
* the complexity of the work.

According to the International Construction Costs Survey 2012, the following were average costs per hour:

* Group 1 tradespeople – plumbers, electricians etc. – £30
* Group 2 tradespeople – carpenters, bricklayers etc. – £30
* Group 3 tradespeople – tillers, carpet layers and plasterers – £30
* General labourers – £18
* Site supervisors – £46.

Quotes, estimated prices and tenders

As we have already seen, estimates, quotes and tenders are very different. It is useful to look at these in slightly more detail, as can be seen in Table 2.7 below.

Type of costing	Explanation
Estimate	This needs to be a realistic proposal of how much a job will cost. An estimate is not binding and the client needs to understand that the final cost might be more.
Quote	This is a fixed price based on a fixed specification. The final price may be different if the fixed specification changes, for example if the customer asks for additional work then the price will be higher.
Tender	This is a competitive process. The customer advertises the fact that they want a job done and invites tenders. The customer will specify the specifications and schedules and may even provide the drawings. The companies tendering then prepare their own documents and submit their price based on the information the customer has given them. All tenders are submitted to the customer by a particular date and are either open or closed. The customer then opens all tenders on a given date and awards the contract to the company of their choice. This process is particularly common among public sector customers, such as local authorities.

Table 2.7 Estimates, quotes and tenders

Implications of inaccurate estimates

Larger companies will have an estimating team. Smaller businesses will have someone who has the job of being an estimator. Whenever they are pricing a job, whether it is a quote, an estimate or a tender, they will have to work out the costs of all materials, labour and other costs. They will also have to include a **mark-up**.

It is vital that all estimating is accurate. Everything needs to be measured and checked. All calculations need to be double-checked.

It can be disastrous if these figures are wrong because:

* if the figure is too high then the client is likely to reject the estimate and look elsewhere as some competitors could be cheaper

* if the figure is too low then the job may not provide the business with sufficient profit and it will be a struggle to make any money out of the job.

KEY TERMS

Mark-up

– a builder or building business, just like any other business, needs to make a profit. Mark-up is the difference between the total cost of the job and the price that the customer is asked to pay for the work.

DID YOU KNOW?

Many businesses fail as a result of not working out their costs properly. They may have plenty of work but they are making very little money.

COMMUNICATING WORKPLACE REQUIREMENTS

Communication can be split into two different types:

* Verbal communication – including face-to-face conversations, discussions in meetings or performance reviews and talking on the telephone.

* Written communication – including all forms of documents, from letters and emails to drawings and work schedules.

Each of these forms of communication needs to be clear, accurate and designed in such a way as to make sure that whoever has to use it or refer to it understands it.

Figure 2.26 It's important to communicate effectively, whether it's verbal or written

Key personnel in the communication cycle

Each construction job will require the services of a team of professionals. They have to be able to work and communicate effectively with one another. Each team has different roles and responsibilities. They can be broken down into three particular groups:

* on site   * off site   * visitors.

These are described in Tables 2.8, 2.9 and 2.10.

Role	Responsibilities
Apprentices	They can work for any of the main building services trades under supervision. They only carry out work that has been specifically assigned to them by a trainer, a skilled operative or a supervisor.
Skilled or trade operative	A specialist in a particular trade, such as bricklaying or carpentry. They will be qualified in that trade, or working towards their qualification
Unskilled operatives	Also known as labourers, these are entry level operatives without any formal training. They may be experienced on sites and will take instructions from the supervisor or site manager.
Building services engineers	They are involved in the design, installation and maintenance of heating, water, electrics, lighting, gas and communications. They work either for the main contractor or the architect and give instruction to building services operatives.
Building services operatives	They include all the main trades involved in installation, maintenance and servicing. They take instruction from the building services engineers and work with other individuals, such as the supervisor and charge-hand.
Charge-hand	This person supervises a specific trade, such as carpenters and bricklayers.
Trade foreperson	This person supervises the day-to-day running of the site, and organises the charge-hand and any other operatives.
Site manager	This person runs the construction site, makes plans to avoid problems and meet deadlines, and ensures all processes are carried out safely. They communicate directly with the client.
Supervisor	The supervisor works directly for the site manager on larger projects and carries out some of the site manager's duties on their behalf.
Health and safety officer	This person is responsible for managing the safety and welfare of the construction site. They will carry out inspections, provide training and correct hazards.

Table 2.8 On-site construction team

Role	Responsibilities
Client	The client, such as a local authority, commissions the job. They define the scope of the work and agree on the timescale and schedule of payments.
Customer	For domestic dwellings, the customer may be the same as the client, but for larger projects a customer may be the end user of the building, such as a tenant renting local authority housing or a business renting an office. These individuals are most affected by any work on site. They should be considered and informed with a view to them suffering as little disruption as possible.
Architect	They are involved in designing new buildings, extensions and alterations. They work closely with clients and customers to ensure the designs match their needs. They also work closely with other construction professionals, such as surveyors and engineers.
Consultant	Consultants such as civil engineers work with clients to plan, manage, design or supervise construction projects. There are many different types of consultant, all with particular specialisms.
Main contractor	This is the main business or organisation employed to head up the construction work. The contractor organises the on-site building team and pulls together all necessary expertise. They manage the whole project, taking full responsibility for its progress and costs.
Clerk of works	This person is employed by the architect on behalf of a client. They oversee the construction work and ensure that it represents the interests of the client and follows agreed specifications and designs.
Quantity surveyor	Quantity surveyors are concerned with building costs. They balance maintaining standards and quality against minimising the costs of any project. They need to make choices in line with Building Regulations. They may work either for the client or for the contractor.
Estimator	Estimators calculate detailed cost breakdowns of work based on specifications provided by the architect and main contractor. They work out the quantity and costs of all building materials, plant required and labour costs.
Sub-contractor	They carry out work on behalf of the main contractor and are usually specialist tradespeople or professionals, such as electricians. Essentially, they provide a service and are contracted to complete their part of the project.
Supplier/wholesaler contracts manager	They work for materials suppliers or stockists, providing materials that match required specifications. They agree prices and delivery dates.

Table 2.9 Off-site construction team

Site visitor	Role and responsibility
Training officers and assessors	These people work for approved training providers. They visit the site to observe and talk to apprentices and their mentors or supervisors. They assess apprentices' competence and help them to put together the paperwork needed to show evidence of their skills.
Building control inspector	This person works for the local authority to ensure that the construction work conforms to regulations, particularly the Building Regulations. They check plans, carry out inspections, issue completion certificates, work with architects and engineers and provide technical knowledge on site.
Water inspector	This person carries out checks of plumbing and drainage systems on construction sites.
Health and Safety Executive (HSE) inspector	An HSE inspector can enter any workplace without giving notice. They will look at the workplace, the activities and the management of health and safety to ensure that the site complies with health and safety laws. They can take action if they find there is a risk to health and safety on site.
Electrical services inspector	Inspectors are approved by the National Inspection Council for Electrical Installation Contracting. They check all electrical installation has been carried out in accordance with legislation, particularly Part P of the Building Regulations.

Table 2.10 Construction visitors

Effects of poor communication

Effective communication is essential in all types of work. It needs to be clear and to the point, as well as accurate. Above all it needs to be a two-way process. This means that any communication that you have with anyone must be understood by them. It means thinking before communicating. Never assume that someone understands you unless they have confirmed that they do.

In construction work you have to keep to schedule and work on time, and it is important to follow precise instructions and specifications. Failing to communicate will always cause confusion, extra cost and delays; it can lead to problems with health and safety and accidents. Such problems are unacceptable and very easy to avoid. Negative communication or poor communication can damage the confidence that others have in you to do your job.

Good communication means efficiency and achievement.

REED TIP...

As well as within your own team, it is important to communicate clearly with the other trades working on a site, especially if there's a problem that may delay the next stage of the job.

Communication techniques and teamwork

It is important to have a good working relationship with colleagues at work. An important part of this is to communicate in a clear way with them. This helps everyone understand what is going on and what decisions have been made. It also means being clear. Most communication with colleagues will be verbal (spoken). Good communication means:

* cutting out mistakes and stoppages (saving money)

* avoiding delays

* making sure that the job is done right the first time and every time.

Figure 2.27 A water inspection

Equality and diversity in communication

Equality and diversity is not simply about treating everyone in the same way. It is actually recognising that people are different and have different needs. Each of us is unique. This could mean that you are working with people of a different culture, a different age (younger or older), or who follow different religions. It might refer to marital status or gender, sexual orientation or your first language.

In all your actions and your communications you should:

* recognise and respect other people's backgrounds

* recognise that everyone has rights and responsibilities

* not harass or be offensive and use language or behaviour that discriminates.

You should also remember that not everyone's first language will be English so they may not understand everything or be able to communicate clearly with you. You might also find that some colleagues may have hearing impairments (or may not hear what you're saying because they are in a noisy environment). In cases like these, use simple language and check that both you and the person you are communicating with have understood the message.

CASE STUDY

South Tyneside Homes

South Tyneside Council's Housing Company

Using writing and maths in the real world

Gary Kirsop, Head of Property Services, says:

'People seem to think that trades are all about your hands, but it's more than that. You're measuring complicated things – all the trades need to have about the same technical level for planning, calculation and writing reports. You need that level to get through your exams for the future too. When you have one day a week in college, but four days a week working with customers in the real world, without communications skills, it would all fall apart. You have to understand that people come from different backgrounds and that they have their own communication modes. Having good GCSEs will really help you get by in the trade.'

Advantages and disadvantages of different methods of communication

As you progress in your career in construction, you may come across a number of different documents that are used either in the workplace or are provided to customers or clients. All of these documents have a specific purpose. Their exact design may vary from business to business, but the information contained on them will usually be similar.

Documents in the workplace
This group of documents tend to be used only within the workplace. Their general purpose is to collect information or to pass on information from one part of the business to another.

Document type	Purpose
Job specifications	These are detailed sets of requirements that cover the construction, features, materials, finishes and performance specifications required for each major aspect of a project. They may, for example, require a particular level of energy efficiency.
Plans or drawings	These are prepared by architects. They are drawn to scale and provide a standard detailed drawing. They will be used as blueprints (instructions) by building services engineers and operatives while they are working on the site.
Work programmes	These are detailed breakdowns of the order in which work needs to be completed, along with an estimate as to how long each stage is likely to take. For example, a certain amount of time will be allocated for site preparation and then piling and the construction of the substructure of the building. The work programme will indicate when particular skills will be needed and for approximately how long.
Purchase orders	These are documents issued by the buyer to a supplier. They detail the type of materials, quantity and the agreed price so form a record of what has been agreed. The order for materials will have been discussed with the supplier before the purchase order is completed. Many purchase orders are now transmitted electronically, although paper records may be necessary for future reference.
Delivery notes	These are issued to the buyer by the supplier. They act as a checklist for the buyer to ensure that every item requested on the purchase order has been delivered. The buyer will sign the delivery note when they are satisfied with the delivery.
Timesheets	These are completed by those working on site and are verified by the charge-hand, site manager or supervisor. They detail the start and finish times of each individual working on site. They form the basis of the pay calculation for that worker and the overall time that the job has taken.
Policy documents	These cover health and safety, environmental or customer service issues, among others. They outline the requirements of all those working on the site. They will identify roles and responsibilities, codes of conduct or practice, and methods and remedies for dealing with problems or breaches of policy.

Table 2.11 Documents used in the workplace

Documents for customers and clients

Some documents need to be provided to customers and clients. They are necessary to pass on information and can include records of costs and charges that the customer or client is expected to pay for work carried out. Table 2.12 describes what these documents are and their purpose.

Document	Purpose
Quotations and tenders	Quotations provide written details of the costs of carrying out a particular job. They are based on the specification or requirements of the customer or client. They will usually be written by the main contractor on larger sites. A tender is usually a sealed quotation submitted by a contractor at the same time as tenders from other firms in the hope that their quotation will not only match the requirements but will also be the cheapest and therefore the most likely to win the work.
Estimates	An estimate differs from a quotation because it is not a binding quote but a calculation of the cost based on what the contractor thinks the work may involve.
Invoices	An invoice is a list of materials or services that have been provided. Each has an itemised cost and the total is shown at the bottom of the document, along with any additional charges such as VAT.

Document	Purpose
Account statements	This is a record of all the transactions (invoices and payments) made by a customer or client over a given period. It matches payments by the customer and client against invoices raised by the supplier. It also notes any money still owing or over-payments that may have been made.
Contracts	A contract is a legally binding agreement, usually between a contractor and a customer or client, which states the obligations of both parties. A series of agreements are made as part of the contract. It binds both parties to stick to the agreement, which may detail timescales, level of work or costs.
Contract variations	Contract variations are also legally binding. They may be required if both the supplier and the customer or client agrees to change some of the terms of the original contract. This could mean, for example, additional obligations, renegotiating prices or new timescales.
Handover information	Once a project, such as an installation, has been completed, the installer that commissioned the installation will check that it is performing as expected. Handover information includes: • the commissioning document, which details the performance and the checks or inspections that have been made • an installation certificate, which shows that the work has been carried out in accordance with legal requirements and the manufacturer's recommendations.

Table 2.12 Documents used with customers and clients

Other forms of communication

So far we have mainly focused on written forms of communication.

KEY TERMS

VAT

– Value Added Tax is charged on most goods and services. It is charged by businesses or individuals that have raised invoices in excess of £73,000 per year and is currently 20 per cent of the bill.

ITEM	DESCRIPTION	QUANTITY	UNIT	RATE £	AMOUNT £	
	Superstructure: Suspended upper floor					
A	Supply and fit the following C16 grade preservative treated softwood					
A1	50 × 195 mm joists	250	m	6.44	1610	00
A2	75 × 195 mm joists	60	m	8.58	514	80
A3	38 × 150 mm strutting	70	m	3.85	339	50
	Carried to collection:			£	2464	30

Figure 2.28 An example bill of quantities

However one of the most common forms of communication is the telephone, whether landline or mobile. The key advantage of a conversation is that problems and queries can be immediately sorted out. However the biggest problem is that there is no record of any decisions that have been made. It is therefore often wise to ask for written confirmation of anything that has been agreed, perhaps in the form of an email.

Construction is one of the many industries that still prefer to have hard copies of documents. It has been made much easier to send copies of documents as email attachments. The problem though is having an available printer of sufficient quality and size to print off attached documents.

Performance reviews

As you progress through your construction career you will be expected to attend performance reviews. This is another form of communication between you and your immediate supervisor. Certain levels of performance will be expected and will have been agreed at previous reviews. At each review your performance, compared to those standards, will be examined. It gives both sides an opportunity to look at progress. It can help identify areas where you might need additional training or support. It may also show areas of your work that need improvement and more effort from you.

Meetings

Meetings also offer important opportunities for communication. They are usually quite structured and will have a series of topics that form what is known as an **agenda**.

Meetings should give everyone the opportunity to contribute and make suggestions as to how to go forward on particular projects and deal with problems. Individuals are often given the job of preparing information for meetings and then presenting it for discussion. A disadvantage of meetings is that while they are happening construction work is not taking place. This means that it is important to run meetings efficiently and not waste time – but also to ensure that everything that needs to be discussed is covered so that extra meetings do not have to be arranged.

Letters

Today emails have largely overtaken more traditional forms of communication, but letters can still be important. Letters obviously need to be delivered so take longer to arrive than emails but sometimes things do need to be sent through the post. It is polite to put in a **covering letter** with documents or other written communication with clients.

Signs and posters

On a daily basis you will also see a range of signs and posters around larger construction sites. Signs are used to communicate either warnings or information and a full list of different types of sign, particularly those relating to health and safety, can be seen in Chapter 1. Their purpose is to be clear and informative. Posters are often put up in communal areas, such as where you might have lunch or keep your personal belongings. These are designed to be simple and to give you vital information. One disadvantage of signs and posters is that they are a one-way form of communication so if you need more information about them you will need to speak to your supervisor.

REED TIP

A good supervisor will make sure you understand what is expected of you in terms of quality, quantity, the speed of the work and how you'll be working with other trades.

KEY TERMS

Agenda

– a brief list of topics to be discussed at a meeting, outlining any decisions that need to be made.

Covering letter

– this is a very brief letter, often just one paragraph long, which states the purpose of the communication and lists any other documents that have been included.

TEST YOURSELF

1. If a drawing is at a scale of 1:500, each millimetre in the drawing represents how much on the ground?

   a. 1 m

   b. 500 cm

   c. 500 mm

   d. 500 m

2. What is the other term used to describe an orthographic projection?

   a. First angle

   b. Second angle

   c. Assembly drawing

   d. Isometric

3. Which of the following are examples of a manufacturer's technical information?

   a. Data sheets

   b. User instructions

   c. Catalogues

   d. All of these

4. On a drawing, if you were to see the letters FDN, what would that mean?

   a. The signature of the architect

   b. Foundation Design Network

   c. Foundations

   d. Full distance

5. If a drawing is at a scale of 1:5, how many times smaller is the drawing than the real object?

   a. 5 times

   b. 50 times

   c. Half the size

   d. 500 times

6. Which of the following values is pi?

   a. 3.121

   b. 3.424

   c. 3.142

   d. 3.421

7. Which document is used to give detailed sets of requirements that cover the construction, features, materials and finishes?

   a. Work programme

   b. Purchase order

   c. Policy document

   d. Job specification

8. What is VAT?

   a. Volume Added Turnover

   b. Vehicle Attendance Tax

   c. Voluntary Aided Trading

   d. Value Added Tax

9. Which individual on a typical site would sign off timesheets?

   a. Architect

   b. Site manager/supervisor

   c. Delivery driver

   d. Customer

10. Which are the two main types of communication?

    a. Verbal and written

    b. Telephones and emails

    c. Meetings and memorandum

    d. Plans and faxes

Chapter 3

Unit CSA–L2Core05

UNDERSTANDING CONSTRUCTION TECHNOLOGY

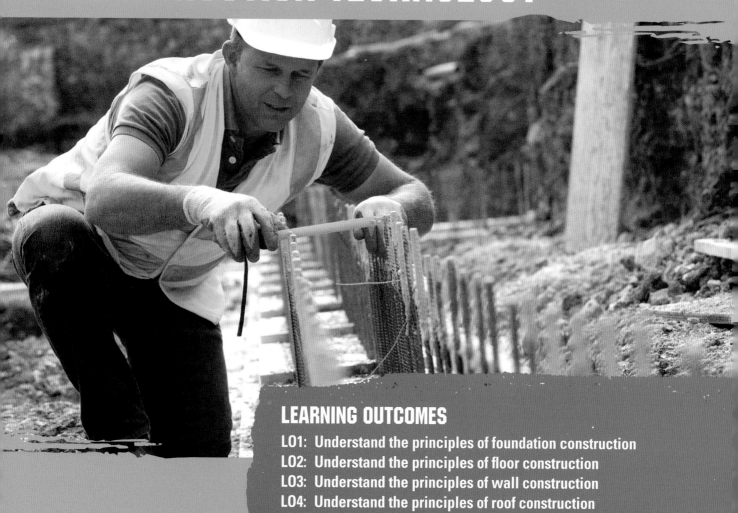

LEARNING OUTCOMES

LO1: Understand the principles of foundation construction

LO2: Understand the principles of floor construction

LO3: Understand the principles of wall construction

LO4: Understand the principles of roof construction

LO5: Understand the supply of utilities and services within construction

LO6: Understand the principle of sustainability within construction

INTRODUCTION

The aim of this chapter is to:

* help you understand the range of building materials used within the construction industry

* help you understand their suitability in the construction of modern buildings.

FOUNDATION CONSTRUCTION

Foundations are the primary element of a building as they support and protect the superstructure (the visible part of the building) above. Foundations are part of the substructure of the building, meaning that they are not visible once the building has been completed.

Foundations spread the load of the superstructure and transfer it to the ground below. They provide the building with structural stability and help to protect the building from any ground movement.

Purpose of foundations

It is important to work out the necessary width of foundations. This depends on the total load of the structure and the load-bearing capacity of the ground or subsoil on which the building is being constructed. This means:

* wide foundations are used when the construction is on weak ground, or the superstructure will be heavy

* narrow foundations are used when the subsoil is capable of carrying a heavy weight, or the building is a relatively light load.

The load that is placed on the foundations spreads into the ground at 45°. **Shear failure** will take place if the thickness of the foundations (T) is less than the projection of the wall or column face on the edge of the foundations (P). This is what leads to subsidence (the ground under the structure sinking or collapsing).

As we will see in this section, the depth of the foundation is dependent on the load-bearing capacity of the subsoil. But for the most part foundations should be 200 mm to 300 mm thick.

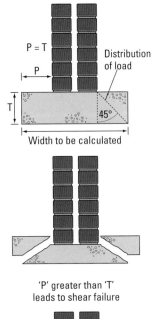

$P = T$

Distribution of load

P

T

45°

Width to be calculated

'P' greater than 'T' leads to shear failure

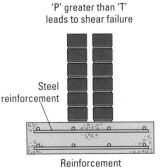

Steel reinforcement

Reinforcement used to reduce 'T'

Figure 3.1 Foundation properties

Different types of foundation

The traditional **strip foundation** is quite narrow and tends to be used for low-rise buildings and dwellings. Most buildings have had unreinforced strip foundations and they were constructed with either brick or block masonry up to the damp course level. Strip foundations can be stepped on sloping ground, in order to cut down on the amount of excavation needed. In poor soil conditions, deep strip foundations can also be used.

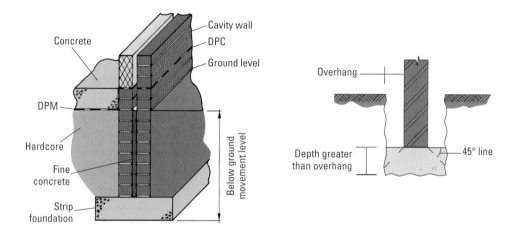

Figure 3.2 Unreinforced strip foundation

Narrow deep strip or trench fill foundations are dug to the foundation depth and then filled with concrete. This reduces excavation, as no bricks or blocks have to be laid into the trench. Trench fill:

* reduces the need to have a wide foundation

* reduces construction time

* speeds up the construction of the footings.

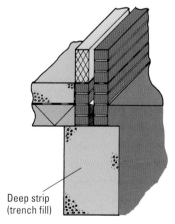

Figure 3.3 Trench fill foundations

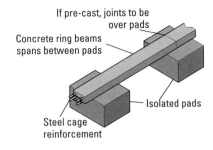

If pre-cast, joints to be over pads

Concrete ring beams spans between pads

Isolated pads

Steel cage reinforcement

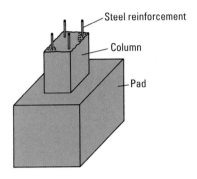

Steel reinforcement

Column

Pad

Figure 3.4 Pad foundations

Pad foundations tend to be used for structures that have either a concrete or a steel frame. The pads are placed to support the columns, which transfer the load of the building into the subsoil.

Pile foundations tend to be used for high-rise buildings or where the subsoil is unstable. Holes are bored into the ground and filled with concrete or pre-cast concrete, steel or timber posts are driven into the ground. These piles are then spanned with concrete ring beams with steel reinforcement so that the load of the building is transferred deeper into the ground below. Pile foundations can be short or long depending on how high the building is or how bad the soil conditions are.

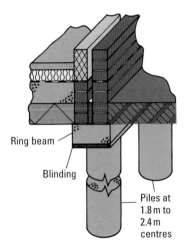

Ring beam

Blinding

Piles at 1.8 m to 2.4 m centres

Figure 3.5 Pile foundations

Raft foundations are used when there is a danger that the subsoil is unstable. A large concrete slab reinforced with steel bars is used to outline the whole footprint of the building. It has an edge beam to take the load from the walls, which is transferred over the whole raft. This means that the building effectively 'floats' on the ground surface on top of the concrete raft.

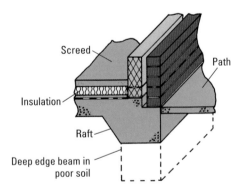

Screed

Path

Insulation

Raft

Deep edge beam in poor soil

Figure 3.6 Raft foundations

Selecting a foundation

One of the first things that a structural engineer will look at when they investigate a site is the nature of the soil and issues such as the water table (where groundwater begins). The type of soil or ground conditions are be very important, as is the possibility of ground movement.

Table 3.1 shows different types of subsoil and how they can affect the choice of foundation.

Subsoil type	Characteristics
Rock	High load bearing but there may be cracks or faults in the rock, which could collapse.
Granular	Medium to high load bearing and can be compacted sand or gravel. If there is a danger of flooding the sand can be washed away.
Cohesive	Low to medium load bearing, such as clay and silt. These are relatively stable, but may have problems with water.
Organic	Low load bearing, such as peat and topsoil. Organic material must be removed before starting the foundations. There is also a great deal of air and water present in the soil.

Table 3.1 Different types of subsoil

The ground may move, particularly if the conditions are wet, extremely dry or there are extremes of temperature. Clay, for example, will shrink in the hot summer months and swell up again in the wet winter months. Frost can affect the water in the ground, causing it to expand.

Ground movement is also affected by the proximity of trees and large shrubs. They will absorb water from the soil, which can dry out the subsoil. This causes the soil underneath the foundations to collapse.

The end use of the building

The other key factor when selecting a foundation is the end use of the building:

* Strip foundation – this is the most common and cheapest type of foundation. Strip foundations are used for low to medium rise domestic and industrial buildings, as the load-bearing will not be high.

* Raft foundation – this is only ever really used when the ground on which the building is being constructed is very soft. It is also sometimes used when the ground across the area is likely to react in different ways because of the weight of the building. In areas of the UK where there has been mining, for example, raft foundations are quite common, as the building could subside. The raft is a rigid, concrete slab reinforced with steel bars. The load of the building is spread across the whole area of the raft.

* Piled foundations – these are used for high-rise buildings where the building will have a high load or where the soil is found to be poor.

Materials used in the construction of foundations

Concrete

Concrete is used to produce a strong and durable foundation. The concrete needs to be poured into the foundation with some care. The size of the foundation will usually determine whether the concrete is actually mixed on site or brought in, in a ready-mixed state, from a supplier. For smaller foundations a concrete mixer and wheelbarrows are usually sufficient. The concrete is then poured into the foundation using a chute.

Concrete consists of both fine and coarse aggregate, along with water, cement and additives if required.

Aggregates

Aggregates are basically fillers. The coarse aggregate is usually either crushed rock or gravel. The grains are 5 mm or larger.

Fine aggregate is usually sand that has grains smaller than 5 mm.

The fine aggregate fills up any gaps between the particles in the coarse aggregate.

Cement

Cement is an adhesive or binder. It is Portland stone, crushed, burnt and crushed again and mixed with limestone. The materials are powdered and then mixed together to create a fine powder, which is then fired in a kiln.

Water

Potable water, which is water that is suitable for drinking, should be used when making concrete. The reason for this is that drinkable water has not been contaminated and it does not have organic material in it that could rot and cause the concrete to crack. The water mixes with the cement and then coats the aggregate. This effectively bonds everything together.

Additives

Additives, or admixtures, make it possible to control the setting time and other aspects of fresh concrete. It allows you to have greater control over the concrete. Common add mixtures can accelerate the setting time, or reduce the amount of water required. They can:

* give you higher strength concrete

* provide protection against corrosion

* accelerate the time the concrete needs to set

* reduce the speed at which the concrete sets

* provide protection against cracking as the concrete sets (prevent shrinkage)

* improve the flow of the concrete

* improve the finish of the concrete

* provide hot or cold weather protection (a drop or rise in temperature can change the amount of time that a concrete needs to set, so these add mixtures compensate for that).

Reinforcement

Steel bars or mesh can be used to give the foundation additional strength and support. It can also stop the foundation from cracking. Concrete is very good at dealing with loads, so weight coming from above is something concrete can deal with. But when concrete foundations are wide, and parts of them are under tension, there is a danger it may crack.

Concrete should also be levelled, usually with a vibrator or a compactor, although newer types of concrete are self-compacting. All concrete needs to be laid on well-compacted ground.

Figure 3.7 Reinforcement using steel bars (or mesh)

FLOOR CONSTRUCTION

For most domestic buildings floorboards or sheets are laid over timber joists. In other cases, and in most industrial buildings, the ground floors have a block and beam construction with hard core. They then have a damp-proof membrane and over the top is solid concrete.

A floor is a level surface that provides some insulation and carries any loads (for example furniture) and to transfer those loads.

The ground floors have additional purposes. They must stop moisture from entering the building from the ground. They also need to prevent plant or tree roots from entering the building.

Ground floors

For ground floors there are two options:

* Solid – is in contact with the ground.

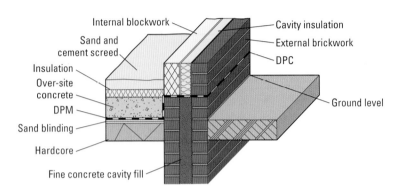

Figure 3.8 Solid ground floors

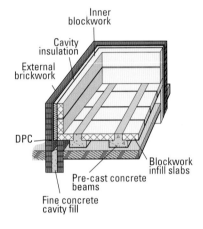

Figure 3.9 Suspended ground floors

* Suspended – the floor does not touch the ground and spans between walls in the building. Effectively there is a void beneath the floor, with air bricks in external walls to allow for ventilation.

The options for ground floors are more complicated than those for upper floors. This is because the ground floors need to perform several functions. It is quite rare for modern buildings to have timber joists and floorboards. Suspended ground floors and traditional timber floors tend to be seen in older buildings. It is far more common to have solid ground floors, or to have timber floors over concrete floors, which are known as floating ground floors.

The key options are outlined in Table 3.2.

Type of floor	Construction and characteristics
Solid	One construction method is to use hard core as the base, with a layer of sand and a layer of insulation such as Celotex, usually 100 mm thick, and then covered with a damp-proof membrane. The concrete is then poured into the foundation. To provide a smooth finish for floor finishes a cement and sand screed is applied, usually after the building has been made watertight..
Timber suspended	A similar process to a solid ground floor is carried out but then, on top of this, dwarf or sleeper walls are built. These are used to support the timber floor. Air bricks are also added to provide necessary ventilation. Joists are then spaced out along the dwarf walls. A damp-proof course is inserted under the floor joists and then floorboards or sheets placed on top of the joists.
Beam and block suspended	Concrete beams and lightweight concrete slabs or blocks are used to create the basic flooring. The beams are evenly spaced across the foundation and gaps between the beams are filled with blocks to form the floor. The blocks and beams are then insulated and it is finished off with either a cement screed or a timber floating floor.
Floating	This timber construction goes over the top of concrete floors. Bearers are put down and then the boarding or sheets are fixed to the bearers. The weight of the boards themselves hold them in place.

Table 3.2 Construction of ground floors

Upper floors

Timber is usually used for these suspended floors in homes and other types of dwellings. In industrial buildings concrete tends to be used.

Timber suspended upper floor

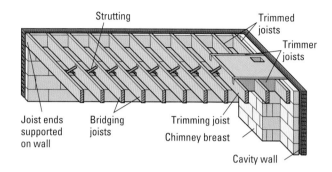

Concrete suspended upper floor

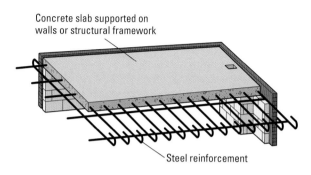

Figure 3.10 Upper floors

For dwellings, bridging joists are used. These are supported at their ends by load-bearing walls. Boarding or sheets provide the flooring for the room on the top of the joists. Underneath the joists plasterboard creates the basis of the ceiling for the room below.

It is also possible to fill the voids between the floorboards and the plasterboard with insulation. Insulation not only helps to prevent heat loss, but can also reduce noise.

Concrete suspended floors are usually either cast on site or available as ready-cast units. They are effectively locked into the structure of the building by steel reinforcement. If the concrete floors are being cast on site then **formwork** is needed. Concrete floors tend to be used in many modern buildings, particularly industrial ones, as they offer greater load bearing capacity, have greater fire resistance and are more sound resistant.

KEY TERMS

Formwork

– this can also be known as shuttering. It is a temporary structure or mould that supports and shapes wet concrete until it cures and is able to be self-supporting.

WALL CONSTRUCTION

Walls have a number of different purposes:

* They hold up the roof.

* They provide protection against the elements.

* They keep the occupants of the building warm.

Many buildings now have double walls, which means as follows:

* The outside wall is a wet one because it is exposed to the elements outside the building.

* The internal wall is dry but it needs to be kept separate from the outside wall by a cavity.

* The cavity or gap acts as a barrier against damp and also provides some heat insulation.

* The cavity can be completely filled or part-filled depending on the insulation value required by Building Regulations.

Within the building there are other walls. These internal walls divide up the space within the building. These do not have to cope with all of the demands of the external walls. As a result, they do not necessarily have to be insulated and are, therefore, thinner. They are block and then covered with plaster. Alternatively they can be a timber framework, which is also known as stud work, and again can be covered with plasterboard.

Different types of wall construction and structural considerations

In addition to walls being external or internal, they can also be classed as being load bearing or non-load bearing.

Internal walls can be either load bearing or non-load bearing. In both internal and external walls, where they are load bearing, any gaps or openings for windows or doors have to be bridged. This is achieved by using either arches or lintels. These support the weight of the wall above the opening.

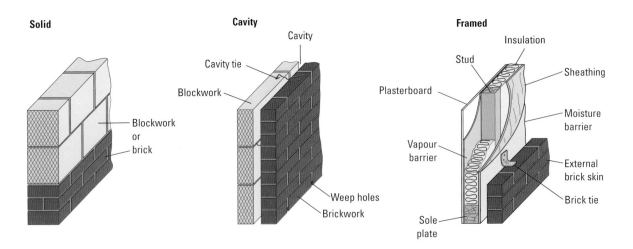

Figure 3.11 Some examples of external wall construction

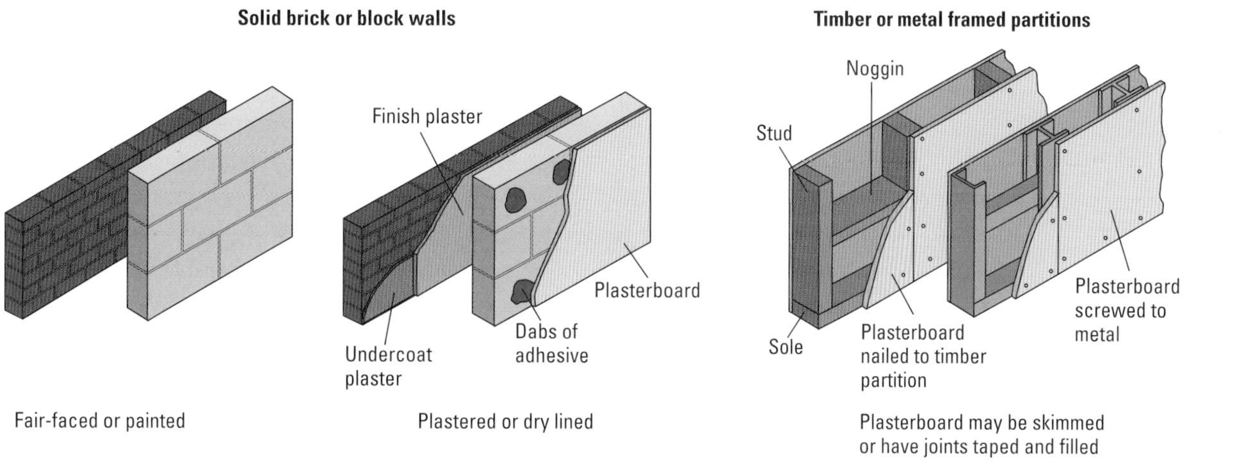

Solid brick or block walls

Timber or metal framed partitions

Finish plaster

Plasterboard

Undercoat plaster

Dabs of adhesive

Fair-faced or painted

Plastered or dry lined

Noggin

Stud

Sole

Plasterboard nailed to timber partition

Plasterboard screwed to metal

Plasterboard may be skimmed or have joints taped and filled

Figure 3.12 Some examples of internal wall construction

Solid masonry

In modern builds solid masonry is quite rare, as it uses up a lot of bricks and blocks. External solid walls tend to be much thinner and made from lightweight blocks in modern builds. They will have some kind of waterproof surface over the top of them, which can be made of render, plastic, metal or timber.

Cavity masonry

As we have seen, cavity walls have an outer and an inner wall and a cavity between them. Usually solid walling, or blockwork, is built up to ground level and then the cavity walling continues to the full height of the building. These cavity walls are ideal for most buildings up to medium height.

Many industrial buildings have cavity walls for the lower part of the building and then have insulated steel panels for the top part of the building.

The usual technique is to have brick for the outer wall and an insulating block for the inner wall. The gap or cavity can then be filled with an insulation material.

Timber framed

Panels made of timber, or in some cases steel, are used to construct walls. They can either be load bearing or non-load bearing and can also be used for the outside of the building or for internal walling. The panels are solid structures and the spaces between the vertical struts (studs) and the horizontal struts (head or sole plates) can be filled with insulation material.

Internal walls or partitions

Internal walls tend to be either solid or framed. Solid walls are made up from blocks. In many industrial buildings the blocks are actually exposed and can be left in their natural state or painted. In domestic buildings plasterboard is usually bonded to the surface and then plastered over to provide a smoother finish.

It is more common for domestic buildings to have framed internal walls, which are known as stud partitions. These are exactly the same as other framed walling, but will usually have plasterboard fixed to them. They would then receive a skimmed coat of plaster to provide the smooth finish.

Damp-proof membrane (DPM) and damp-proof course (DPC)

Damp-proof membranes are installed under the concrete in ground floors in order to ensure that ground moisture does not enter the building. Effectively the membrane waterproofs the building.

Damp-proof courses are a continuation of the damp-proof membrane. They are built into a horizontal course of either block and brickwork, which is a minimum of 150 mm above the exterior ground level. DPCs are also designed to stop moisture from coming up from the ground, entering the wall and then getting into the building. The most common DPC is a polythene sheet called visqueen DPC. It comes in rolls to the appropriate width for the wall.

In older buildings lead, bitumen or slate would have been used as a DPC.

ROOF CONSTRUCTION

In a country such as the UK, with a great deal of rain and sometimes snowy weather, it makes sense for roofs to be pitched. Pitched means built at an angle. The idea is that the rain and snow falls down the angle and off the edge of the roof or into gutters rather than lying on the roof.

However, not all roofs are pitched. In fact many domestic dwelling extensions have flat roofs. A great number of industrial buildings have entirely flat roofs. The problem with a flat roof is that it needs to be able to support itself, but just as importantly it needs to be able to carry the additional weight of snow or rain. This means that large flat roofs may have to have steel sections (known as trusses) or even reinforced concrete and beams to increase their load-bearing capacity.

Roofs also provide stability to the walls by tying them together. As we will see, there are several different types of roof. These are usually identified by their pitch or shape.

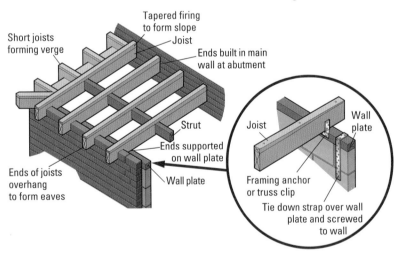

Figure 3.13 Flat roof structure

Types of roof construction

The roof is made up of the rafters and beams. Everything above the framework is regarded as a roof covering, such as slates, tiles and felt.

Table 3.3 outlines some of the key characteristics of different types of roof.

Roof type	Characteristics	How it looks
Flat	This is a roof that has a slope of less than 10°. Generally flat roofs are used for smaller extensions to dwellings and on garages. Traditionally they would have had bitumen felt, although it is becoming more common for fibreglass to be used.	 Figure 3.14
Mono-pitch	This is a roof that has a single sloping surface but is not fixed to another building or wall. The front and back walls could be different heights, or the other exposed surface of the roof is **perpendicular**.	 Figure 3.15
Gambrel roof	This is a roof that has two differently angled slopes. Usually the upper part of the roof has a fairly shallow pitch or slope and the lower part of the roof has a steeper slope.	 Figure 3.16
Couple roof	This is often called gable end and is one of the most common types of roof for dwellings. A gable is a wall with a triangular upper part. This supports the roof in construction using purlins. This means that the roof has two sloping surfaces, which come down from the ridge to the eaves.	 Figure 3.17
Hipped roof	Hipped roofs have slopes on three or four four sides. There are also hipped roofs with single, straight gables.	 Figure 3.18
Lean-to	A lean to is similar to a mono-pitched roof except it is abutted to a wall. The slope is greater than 10°. The higher part of the roof is fixed to a higher wall.	 Figure 3.19

Table 3.3 Different types of roof

Roofing components

Each visible part of a roof has a specific name and purpose. Table 3.4 explains each of these individual features.

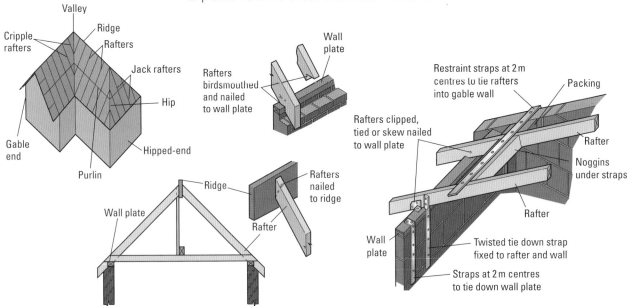

Figure 3.20 Traditional cut roof details

Roof feature	Description
Ridge	This is the top of the roof and the junction of the sloping sides. It is the peak, where the rafters meet.
Purlin	This is a beam that supports the mid-span section of rafters.
Firings	These are angled pieces of timber that are placed on the rafters to create a slope.
Batten	Roof battens are thin strips, usually of wood, which provide a fixing point for either roofing sheets or roof tiles.
Tile	These can be made from clay, slate, concrete or plastic. They are placed in regular, overlapping rows and fixed to the battens.
Fascia	This is a horizontal, decorative board. It is usually a wooden board, although it can be PVC. It is fixed to the ends of the rafters at eaves level and is both a decorative feature and a fixing for rainwater goods.
Wall plate	This is a horizontal timber that is placed at the top of a wall at eaves level. It holds the ends of joists or rafters.
Bracings	Roof rafters need to be braced to make them more rigid and stable. These bracings prevent the roof from buckling. Usually there are several braces in a typical roof.
Felt	Roofing felt has two elements – it has a waterproofing agent (bitumen) and what is known as a carrier. The carrier can be either a polyester sheet or a glass fibre sheet. Roofing felt tends to be used for flat roofs and for roofs with a shallow pitch.
Slate	Slate roofing tiles are usually fixed to timber battens with double nails. They have a lifespan of between 80 and 100 years.
Flashings	Wherever there is a joint or angle on a roof, a thin sheet of either lead or another waterproof material is added. In the past this tended always to be lead. Many different types of flashing can now be used but all have the role of preventing water penetrating into joints, such as on abutments to walls and around the chimney stack.
Rafter	Roof rafters are the main structural components of the roof. They are the framework. They rest on supporting walls. The rafters are set at an angle on sloped roofs or horizontal on a flat roof.
Apex	The apex is the highest point of the roof, usually the ridge line.

Roof feature	Description
Soffit	Soffits are the lower part, or overhanging part, of the eaves. In other words they are the underside of the eaves. A flat section of timber or plastic is usually fixed to the soffit to ensure water tightness.
Bargeboard	This is an ornamental feature, which is fixed to the gable end of a roof in order to hide the ends of roof timbers.
Eaves	These are the area found at the foot of the rafter. They are not always visible as they can be flush. In modern construction, the eaves have two parts: the visible eaves projection and the hidden eaves projection.

Table 3.4 Parts of a roof

Roof coverings

There are many different types of materials that can be used to cover the roof. Even tiles and slates come in a wide variety of shapes and sizes, along with colours and different finishes.

In many cases the type of roof covering is determined by the traditional and local styles in the area. Local authorities want roof coverings that are not too far from the common style in the area. This does not stop manufacturers from coming up with new ideas, however, which can add benefits during construction. There is much innovation and labour saving that also helps to minimise build costs.

Affordable clay tiles, for example, make it possible to use traditional materials that had been out of the budget of many construction jobs for a number of years.

The Table 3.5 outlines some of the more common types of roof covering and describes their main characteristics and use.

Roof covering	Description
Felt Figure 3.21	Felt is used as a waterproof barrier. Internal felt is rolled over the top of the rafters. The strips are overlapped to provide a permanent waterproof barrier. They are then battened down and another roof covering, such as slate or tile, placed over the top of them. For flat roofs, felt is used as the external roof covering and is covered in a waterproof material, such as bitumen.
Slate Figure 3.22	Slate is a flat, natural substance, which is laid onto the battens with each slate tile overlapping the top of the slate in the row directly below it. The slate tiles are either nailed or hooked into place.
Tile Figure 3.23	There is a huge variety of roofing tiles, made from clay, ceramics or concrete. They are designed and moulded so that they overlap with one another and are fixed to the roof in a similar way to slate tiles.

Roof covering	Description
Metals  Figure 3.24	There are many different types of metal roof covering, such as corrugated sheets, flat sheets, box profile sheets or even sheets that have a tile effect. The metal is galvanised and plastic coated to provide a durable and long-lasting waterproof surface.

Table 3.5 Roof coverings

CASE STUDY

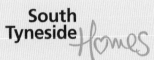

South Tyneside

South Tyneside Council's
Housing Company

How to impress in interviews

Andrea Dickson and Gillian Jenkins sit on the interview panels for apprenticeship applications at South Tyneside Homes.

'Interviews are all about the three Ps: Preparation, Presentation and Personality.

An applicant should turn up with some knowledge about the apprenticeship programme and the company itself. For example, knowing how long it is, that they have to go to college and to work – and don't say, "I was hoping you'd tell me about it"! If they've done a bit of research, it will show through and work in their favour – especially if they can explain why it is that they want to work here.

It sets them up for the interview if they come in smartly dressed. We're not marking them on that, but it does show respect for the situation. It's still a formal process and although we try to make them feel at ease as much as we possibly can, there's no getting away from the fact that they're applying for a job and it is a formal setting.

The interviews are a chance to tell the company about themselves: what they do in their spare time, what their greatest achievements have been and why. Applicants should talk about what interests them, for example, are they really interested in becoming a joiner or is that something their parents want them to do? An apprenticeship has to be something they really want to do – if they have enthusiasm for the programme, then they'll fly through it. If not, it's a very long three to four years. Without that passion for it, the whole process will be a struggle; they'll come in late to work and might even fail exams.

We also talk to them about any customer service experiences they've had, working in a team, project working (for example, a time you had to complete a task and what steps you took), as well as asking some questions about health and safety awareness.'

SUPPLY OF UTILITIES AND SERVICES

Most but not all dwellings and other structures are connected in some way to a wide range of utilities and services. In the majority of cities, towns and villages structures are connected to key utilities and services, such as a sewer system, potable (drinking) water, gas and electricity. This is not always the case for more remote structures.

Whenever construction work is carried out, whether it is on an existing structure or a new build, the supply of utilities and services or the linking up of these parts of the **infrastructure** are very important. Often they will require the services of specialist engineers from the **service provider**.

Table 3.6 outlines the main utilities and services that are provided to most structures.

KEY TERMS

Infrastructure

– these are basic facilities, such as a power supply, a road network and a communication link.

Service provider

– these are companies or organisations that provide utilities, such as gas, water, communications or electricity.

Utility or service	Description
Drainage	Drainage is delivered by a range of water and sewerage companies. They are responsible for ensuring that surface water can drain away into their system.
Waste water and sewerage	Any waste water and sewage generated by the occupants of a structure needs to have the necessary pipework to link it to the main sewerage system. It is then sent to a sewage treatment works via the pipework. Remote areas may not be connected to the sewerage system so use septic tanks and cesspools.
Water	Each structure should be linked to the water supply that provides wholesome, potable drinking water. The pipework linking the structure to the water supply needs to be protected to ensure that backflow from any other source does not contaminate the system.
Gas	Each area has a range of different gas suppliers. This is delivered via a service pipe from the main system into the structure. Areas that do not have access to the main gas supply system use gas contained in bottles.
Electricity	The National Grid provides electricity to a variety of different electricity suppliers. It is the National Grid that operates and maintains the cabling. There are around 28 million individual customers in the UK.
Communications (telephone, data, cable)	There are several ways in which telecommunications can be linked to a structure. Traditional telephone poles hold up copper cables and not only provide telephone but also internet access to structures. In cities and many of the larger towns this system is being replaced by cables that are fibre optic and run underground. These are then linked to each individual structure.
Ducting (heating and ventilation)	Heating and ventilation engineers install and maintain duct work. The complex systems are known as HVAC. These systems can transfer air for heating or cooling of the structure. The overall system can also provide hot and cold water systems, along with ventilation.

Table 3.6 Services and utilities

SUSTAINABILITY AND INCORPORATING SUSTAINABILITY INTO CONSTRUCTION PROJECTS

Sustainability is something that we all need to be concerned about as the earth's resources are used up rapidly and climate change becomes an ever-bigger issue. Carbon is present in all fossil fuels, such as coal or natural gas. Burning fossil fuels releases carbon dioxide, which is a greenhouse gas linked to climate change.

Energy conservation aims to reduce the amount of carbon dioxide in the atmosphere. The idea is to do this by making buildings better insulated and, at the same time, make heating appliances more efficient. Sustainability also means attempting to generate energy using renewable and/or low or zero carbon methods.

According to the government's Environment Agency, sustainable construction means using resources in the most efficient way. It also means cutting down on waste on site and reducing the amount of materials that have to be disposed of and put into **landfill**.

In order to achieve sustainable construction the Environment Agency recommends

* reducing construction, demolition and excavation waste that needs to go to landfill

* cutting back on carbon emissions from construction transport and machinery

* responsibly sourcing materials

* cutting back on the amount of water that is wasted

* making sure construction does not have a negative impact on **biodiversity**.

Sustainable construction and incorporating it into construction projects

In the past buildings were generally constructed as quickly as possible and at the lowest cost. More recently the idea of sustainable construction focuses on ensuring that the building is not only of good quality and that it is affordable, but that it is also efficient in terms of energy use and resources.

Sustainable construction also means having the least negative environmental impact. So this means minimising the use of raw materials, energy, land and water. This is not only during the build but also for the lifetime of the building.

Finite and renewable resources

We all know that resources such as coal and oil will eventually run out. These are examples of finite resources.

Oil, however, is not just used as fuel – it is in plastic, dyes, lubricants and textiles. All of these are used in the construction process.

Renewable resources are those that are produced either by moving water, the sun or the wind. Materials that come from plants, such as biodiesel, or the oils used to make adhesives, are all examples of renewable resources.

Figure 3.25 Most modern new-builds follow sustainable principles

The construction process itself is only part of the problem. It is important to consider the longer-term impact and demands that the building will have on the environment. This is why there has been a drive towards sustainable homes and there is a Code for Sustainable Homes.

Construction and the environment

In 2010, construction, demolition and excavation produced 20 million tonnes of waste that had to go into landfill. The construction industry is also responsible for most illegal fly tipping (illegally dumping waste). In any year the Environment Agency responds to around 350 pollution incidents caused as a result of construction.

Regardless of the size of the construction job, everyone in construction is responsible for the impact they have on the environment. Good site layout, planning and management can help to reduce these problems.

Sustainable construction helps to encourage this because it means managing resources in a more efficient way, reducing waste, recycling where possible and reducing your **carbon footprint**.

Architecture and design

The Code for Sustainable Homes Rating Scheme was introduced in 2007. Many local authorities have instructed their planning departments to encourage sustainable development. This begins with the work of the architect who designs the building.

DID YOU KNOW?

Search on the internet for 'sustainable building' and 'improving energy efficiency' to find out more about the latest technologies and products.

KEY TERMS

Carbon footprint

– This is the amount of carbon dioxide produced by a project. This not only includes burning carbon-based fuels such as petrol, gas, oil or coal, but includes the carbon that is generated in the production of materials and equipment.

Local authorities ask that architects and building designers:

* ensure the land is safe for development – so if it is contaminated this is dealt with first

* ensure access to and protect the natural environment – this supports biodiversity and tries to create open spaces for local people

* reduce the negative impact on the local environment – buildings should keep noise, air, light and water pollution down to a minimum

* conserve natural resources and cut back carbon emissions – this covers energy, materials and water

* ensure comfort and security – good access, close to public transport, safe parking and protection against flooding.

Figure 3.26 Sustainable developments aim to be pleasant places to live

Using locally managed resources

The construction industry imports nearly 6 million cubic metres of sawn wood each year. However there is plenty of scope to use the many millions of cubic metres of timber produced in managed forests, particularly in Scotland.

Local timber can be used for a wide variety of different construction projects:

* Softwood – including pines, firs, larch and spruce – for panels, decking, fencing and internal flooring.

* Hardwood – including oak, chestnut, ash, beech and sycamore – for a wide variety of internal joinery.

Eco-friendly, sustainable manufactured products and environmentally resourced timber

There are now many suppliers that offer sustainable building materials as a green alternative. Tiles, for example, are now made from recycled plastic bottles and stone particles.

There is a National Green Specification database of all environmentally friendly building materials. This provides a checklist where it is possible to compare specifications of sustainable products to traditionally manufactured products, such as bricks.

Simple changes can be made, such as using timber or ethylene-based plastics instead of UPVC window frames, to ensure a building uses more sustainable materials.

As we have seen, finding locally managed resources such as timber makes sense in terms of cost and in terms of protecting the environment. There are many alternatives to traditional resources that could help protect the environment.

The Timber Trade Federation produces a Timber Certification System. This ensures that wood products are labelled to show that they are produced in sustainable forests.

Around 80 per cent of all the softwood used in construction comes from Scandinavia or Russia. Another 15 per cent comes from the rest of Europe, or even North America. The remaining 5 per cent comes from tropical countries, and is usually sourced from sustainable forests.

Alternative methods of building

The most common type of construction is, of course, brick and block work. However there are plenty of other options:

* Timber frame

* Insulated concrete formwork – where a polystyrene mould is filled with reinforced concrete.

* Structural insulated panels – where buildings are made up of rigid building boards, rather like huge sandwiches.

* Modular construction – this uses similar materials and techniques to standard construction, but the units are built off site and transported ready-constructed to their location.

Figure 3.27 Window frames made from timber

DID YOU KNOW?

www.recycledproducts. org.uk has a long list of recycled surfacing products, such as tiles, recycled wood and paving and details of local suppliers.

Figure 3.28 Timber Certification System

Figure 3.29 Green roofing

Figure 3.30 Flooring made from cork

Alternatives to roofing and flooring

There are alternatives to traditional flooring and roofing, all of which are greener and more sustainable. Green roofing has become an increasing trend in recent years. Metal roofs made of steel, aluminium or copper are lightweight and often use a high percentage of recycled metal. Solar roof shingles, or solar roof laminates, while expensive, help to reduce the use of electricity and heating of the dwelling. Some buildings even have a green roof, which consists of a waterproof membrane, a growing medium and plants such as grass or sedum.

Just as roofs are becoming greener, so too are the options for flooring. The use of bamboo, eucalyptus or cork is becoming more common. A new version of linoleum has been developed with **biodegradable**, **organic** ingredients. Some buildings are also using concrete rather than traditional timber floorboards and joists. The concrete can be coloured, stained or patterned.

An increasing trend has been for what is known as off-site manufacture (OSM). European businesses, particularly those in Germany, have built over 100,000 houses. The entire house is manufactured in a factory and then assembled on site. Walls, floors, roofs, windows and doors with built-in electrics and plumbing, all arrive on a lorry. Some manufacturers even offer completely finished dwellings, including carpets and curtains. Many of these modular buildings are actually designed to be far more energy efficient than traditional brick and block constructions. Many come ready fitted with heat pumps, solar panels and triple-glazed windows.

KEY TERMS

Biodegradable

– the material will more easily break down when it is no longer needed. This breaking down process is done by micro-organisms.

Organic

– this is a natural substance, usually extracted from plants.

Energy efficiency and incorporating it into construction projects

Energy efficiency involves using less energy to provide the same level of output. The plan is to try to cut the world's energy needs by 30 per cent before 2050. This means producing more energy efficient buildings. It also means using energy efficient methods to produce materials and resources needed to construct buildings.

Building Regulations

In terms of energy conservation, the most important UK law is the Building Regulations 2010, particularly Part L. The Building Regulations:

* list the minimum efficiency requirements

* provide guidance on compliance, the main testing methods, installation and control

* cover both new dwellings and existing dwellings.

A key part of the regulations is the Standard Assessment Procedure (SAP), which measures or estimates the energy efficiency performance of buildings.

Local planning authorities also now require that all new developments generate at least 10 per cent of their energy from renewable sources. This means that each new project has to be assessed one at a time.

Energy conservation

By law, each local authority is required to reduce carbon dioxide emissions and to encourage the conservation of energy. This means that everyone has a responsibility in some way to conserve energy:

* Clients, along with building designers, are required to include energy efficient technology in the build.

* Contractors and sub-contractors have to follow these design guidelines. They also need to play a role in conserving energy and resources when actually working on site.

* Suppliers of products are required by law to provide information on energy consumption.

In addition, new energy efficiency schemes and building regulations cover the energy performance of buildings. Each new build is required to have an Energy Performance Certificate. This rates a building's energy efficiency from A (which is very efficient) to G (which is least efficient).

Some building designers have also begun to adopt other voluntary ways of attempting to protect the environment. These include BREEAM, which is an environmental assessment method, and the Code for Sustainable Homes, which is a certification of sustainability for new builds.

Figure 3.31 The Energy Saving Trust encourages builders to use less wasteful building techniques and more energy efficient construction

High, low and zero carbon

When we look at energy sources, we consider their environmental impact in terms of how much carbon dioxide they release. Accordingly, energy sources can be split into three different groups:

- high carbon – those that release a lot of carbon dioxide
- low carbon – those that release some carbon dioxide
- zero carbon – those that do not release any carbon dioxide.

Some examples of high carbon, low carbon and zero carbon energy sources are given in Table 3.7 below.

High carbon energy source	Description
Natural gas or LPG	Piped natural gas or liquid petroleum gas stored in bottles
Fuel oils	Domestic fuel oil, such as diesel
Solid fuels	Coal, coke and peat
Electricity	Generated from non-renewable sources, such as coal-fired power stations
Low carbon energy source	
Solar thermal	Panels used to capture energy from the sun to heat water
Solid fuel	Biomass such as logs, wood chips and pellets
Hydrogen fuel cells	Converts chemical energy into electrical energy
Heat pumps	Devices that convert low temperature heat into higher temperature heat
Combined heat and power (CHP)	Generates electricity as well as heat for water and space heating
Combined cooling, heat and power (CCHP)	A variation on CHP that also provides a basic air conditioning system
Zero carbon energy	
Electricity/wind	Uses natural wind resources to generate electrical energy
Electricity/tidal	Uses wave power to generate electrical energy
Hydroelectric	Uses the natural flow of rivers and streams to generate electrical energy
Solar photovoltaic	Uses solar cells to convert light energy from the sun into electricity

Table 3.7 High, low and zero carbon energy sources

It is important to try to conserve non-renewable energy so that there will be sufficient fuel for the future. The fuel has to last as long as is necessary to completely replace it with renewable sources, such as wind or solar energy.

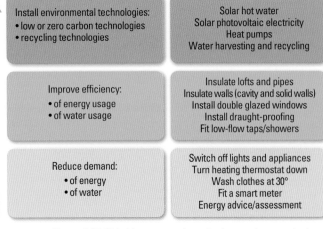

Figure 3.32 Working towards reducing carbon emissions

Alternative energy sources

There are several new ways in which we can harness the power of water, the sun and the wind to provide us with new heating sources. All of these systems are considered to be far more energy efficient than traditional heating systems, which rely on gas, oil, electricity or other fossil fuels.

Solar thermal

At the heart of this system is the solar collector, which is often referred to as a solar panel. The idea is that the collector absorbs energy from the sun, which is then converted into heat. This heat is then applied to the system's heat transfer fluid.

The system uses a differential temperature controller (DTC) that controls the system's circulating pump when solar energy is available and there is a demand for water to be heated.

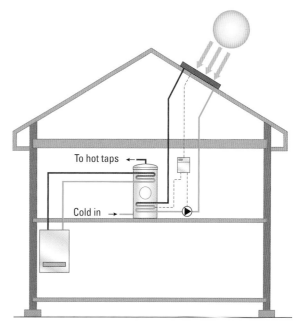

Figure 3.33 Solar thermal hot water system

In the UK, due to the lack of guaranteed solar energy, solar thermal hot water systems often have an auxiliary heat source, such as an immersion heater.

Biomass (solid fuel)

Biomass stoves burn either pellets or logs. Some have integrated hoppers that transfer pellets to the burner. Biomass boilers are available for pellets, woodchips or logs. Most of them have automated systems to clean the heat exchanger surfaces. They can provide heat for domestic hot water and space heating.

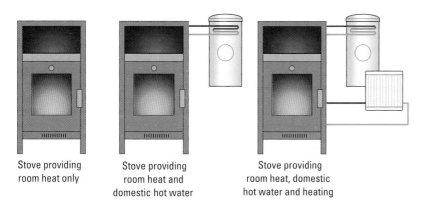

Stove providing room heat only

Stove providing room heat and domestic hot water

Stove providing room heat, domestic hot water and heating

Figure 3.34 Biomass stoves output options

Heat pumps

Heat pumps convert low temperature heat from air, ground or water sources to higher temperature heat. They can be used in ducted air or piped water **heat sink** systems.

KEY TERMS

Heat sink

– this is a heat exchanger that transfers heat from one source into a fluid, such as in refrigeration, air-conditioning or the radiator in a car.

There are different arrangements for each of the three main systems:

* Air source pumps operate at temperatures down to minus 20°C.

* Ground source pumps operate on **geothermal** ground heat.

* Water source systems can be used where there is a suitable water source, such as a pond or lake.

The heat pump system's efficiency relies on the temperature difference between the heat source and the heat sink.

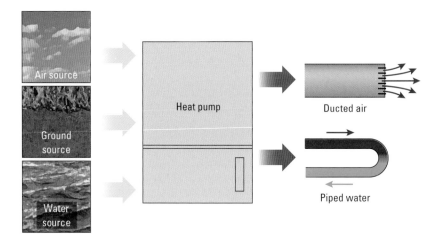

Figure 3.35 Heat pump input and output options

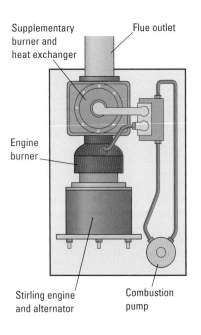

Figure 3.36 Example of a MCHP (micro combined heat and power) unit

Combined heat and power (CHP) and combined cooling heat and power (CCHP) units

These are similar to heating system boilers, but they generate electricity as well as heat for hot water or space heating (or cooling). Electricity is generated along with sufficient energy to heat water and to provide space heating.

Wind turbines

Freestanding or building-mounted wind turbines capture the energy from wind to generate electrical energy. The wind passes across rotor blades of a turbine, which causes the hub to turn. The hub is connected by a shaft to a gearbox. This increases the speed of rotation. A high speed shaft is then connected to a generator that produces the electricity.

Solar photovoltaic systems

A solar photovoltaic system uses solar cells to convert light energy from the sun into electricity.

Energy ratings

Energy rating tables are used to measure the overall efficiency of a dwelling, with rating A being the most energy efficient and rating G the least energy efficient (see Fig 3.41).

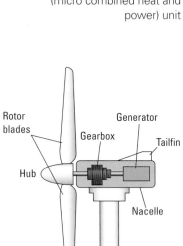

Figure 3.37 A basic horizontal axis wind turbine

Alongside this, there are environmental impact ratings (see Fig 3.40). This type of rating measures the dwelling's impact on the environment in terms of how much carbon dioxide it produces. Again, rating A is the highest, showing it has the least impact on the environment, and rating G is the lowest.

A Standard Assessment Procedure (SAP) is used to place the dwelling on the energy rating table. The ratings are used by local authorities and other groups to assess the energy efficiency of new and old housing and must be provided when houses are sold.

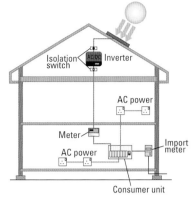

Figure 3.38 A basic solar photovoltaic system

Preventing heat loss

Most old buildings are under-insulated and benefit from additional insulation, whether this is applied to ceilings, walls or floors.

The measurement of heat loss in a building is known as the U Value. It measures how well parts of the building transfer heat. Low U Values represent high levels of insulation. U Values are becoming more important as they form the basis of energy and carbon reduction standards.

By 2016 all new housing is expected to be Net Zero Carbon. This means that the building should not be contributing to climate change.

Many of the guidelines are now part of Building Regulations (Part L). They cover:

* insulation requirements
* openings, such as doors and windows
* solar heating and other heating
* ventilation and air-conditioning
* space heating controls
* lighting efficiency
* air tightness.

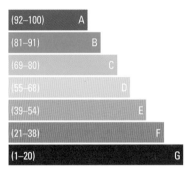

Figure 3.39 SAP energy efficiency rating table – the ranges in brackets show the percentage energy efficiency for each banding

Building design

UK homes spend £2.4bn every year just on lighting. One of the ways of tackling this cost is to use energy saving lights, but also to maximise natural lighting. For the construction industry this means:

* increased window size
* orientating window angles to make the most of sunlight – south facing windows maximise sunlight in winter and limit overheating in the summer
* window design – with a variety of different types of opening to allow ventilation.

Solar tubes are another way of increasing light. These are small domes on the roof, which collect sunlight and then direct it through a tube (which is reflective). It is then directed through a diffuser in the ceiling to spread light into the room.

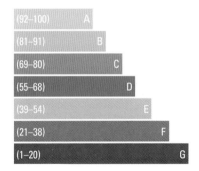

Figure 3.40 SAP environmental impact rating table

TEST YOURSELF

1. In which of the following types of building is a traditional strip foundation used?

   a. High rise

   b. Medium rise

   c. Low rise

   d. Industrial buildings

2. Which of the following is a reason for using a raft foundation?

   a. The subsoil is rock

   b. The subsoil is unstable

   c. The subsoil is stable

   d. The access to the site allows it

3. What holds down a floating floor?

   a. Nails and screws

   b. Adhesives

   c. Blocks

   d. Its own weight

4. What is another term for formwork?

   a. Shuttering

   b. Cavity

   c. Joist

   d. Boarding

5. What is the minimum distance the DPC should be above ground level?

   a. 50 mm

   b. 100 mm

   c. 150 mm

   d. 200 mm

6. A roof is said to be flat if it has a slope of less than how many degrees?

   a. 5

   b. 10

   c. 15

   d. 20

7. What shape is the upper part of a gable end?

   a. Rectangular

   b. Semi-circular

   c. Square

   d. Triangular

8. What do you call the horizontal timber that is placed at the top of a wall at eaves level in a roof, to hold the ends of joists or rafters?

   a. Fascia

   b. Bracings

   c. Wall plate

   d. Batten

9. What happens to the majority of construction demolition and excavation waste?

   a. It is buried on site

   b. It is burned

   c. It goes into landfill

   d. It is recycled

10. Which part of the Building Regulations 2010 requires construction to consider and use energy efficiently?

    a. Part B

    b. Part D

    c. Part K

    d. Part L

INDEX

ACKNOWLEDGEMENTS

The author and the publisher would also like to thank the following for permission to reproduce material:

Images and diagrams

Alamy: Arcaid Images: 3.15, blickwinkel: 3.16, Peter Davey: chapter 1 opener; **BSA**: 2.4; **Energy Saving Trust © 2013**: 3.31; **Fotolia**: 1.1, 1.2, 1.3, 1.5, 1.6, 1.7, 1.8, 1.14, 1.15, 1.16, 2.26; **Helfen**: 2.3; **instant art**: table 1.15; **iStockphoto**: 1.11, 2.27, 3.7, 3.19, 3.21, 3.22, 3.24, 3.25, 3.28, 3.29, 3.3; **Nelson Thornes**: 1.9, 1.10, 1.12, 1.13; **Peter Brett**: 2.2, 2.5, 2.6; **Science Photo Library**: Peter Gardiner: 1.4; **Shutterstock**: chapter 2 and 3 opener, 3.14, 3.17, 3.23, 3.26, 3.27; **Wikipedia**: 3.18.

Every effort has been made to trace the copyright holders but if any have been inadvertently overlooked the publisher will be pleased to make the necessary arrangements at the first opportunity.